T0351140

Security Framework for The Internet of Things Applications

The text highlights a comprehensive survey that focuses on all security aspects and challenges facing the Internet of Things systems, including outsourcing techniques for partial computations on edge or cloud while presenting case studies to map security challenges. It further covers three security aspects including Internet of Things device identification and authentication, network traffic intrusion detection, and executable malware files detection.

This book:

- Presents a security framework model design named Behavioral Network Traffic Identification and Novelty Anomaly Detection for the IoT Infrastructures.
- Highlights recent advancements in machine learning, deep learning, and networking standards to boost Internet of Things security.
- Builds a near real-time solution for identifying Internet of Things devices connecting to a network using their network traffic traces and providing them with sufficient access privileges.
- Develops a robust framework for detecting IoT anomalous network traffic.
- Covers an anti-malware solution for detecting malware targeting embedded devices.

It will serve as an ideal text for senior undergraduate and graduate students, and professionals in the fields of electrical engineering, electronics and communication engineering, computer engineering, and information technology.

Computational Methods for Industrial Applications

Series Editor- Bharat Bhushan

In today's world IoT platforms and processes in conjunction with the disruptive blockchain technology and path breaking AI algorithms lay out a sparking and stimulating foundation for sustaining smarter systems. Further computational intelligence (CI) has gained enormous interests from various quarters in order to solve numerous real-world problems and enable intelligent behavior in changing and complex environment. This book series focuses on varied computational methods incorporated within the system with the help of artificial intelligence, learning methods, analytical reasoning and sense making in big data. Aimed at graduate students, academic researchers and professionals, the proposed series will cover the most efficient and innovative technological solutions for industrial applications and sustainable smart societies in order to alter green power management, effect of carbon emissions, air quality metrics, industrial pollution levels, biodiversity and ecology.

Blockchain for Industry 4.0: Emergence, Challenges, and Opportunities
Anoop V.S, Asharaf S, Justin Goldston and Samson Williams

Intelligent Systems and Machine Learning for Industry: Advancements, Challenges and Practices
P.R Anisha, C. Kishor Kumar Reddy, Nguyen Gia Nhu, Megha Bhushan, Ashok Kumar, Marlia Mohd Hanafiah

Sustainability in Industry 5.0
Theory and Applications

Edited by C Kishor Kumar Reddy, P R Anisha, Samiya Khan, Marlia Mohd Hanafiah, Lavanya Pamulaparty and R Madana Mohana

Security Framework for The Internet of Things Applications

Salma Abdalla Hamad
Quan Z. Sheng
Wei Emma Zhang

CRC Press
Taylor & Francis Group
Boca Raton London New York

CRC Press is an imprint of the
Taylor & Francis Group, an **informa** business
A CHAPMAN & HALL BOOK

Designed cover image: Viewvie/Shutterstock

First edition published 2024
by CRC Press
2385 NW Executive Center Drive, Suite 320, Boca Raton FL 33431

and by CRC Press
4 Park Square, Milton Park, Abingdon, Oxon, OX14 4RN

CRC Press is an imprint of Taylor & Francis Group, LLC

© 2024 Salma Abdalla Hamad, Quan Z. Sheng and Wei Emma Zhang

ISBN: 978-1-032-40927-6 (hbk)
ISBN: 978-1-032-76489-4 (pbk)
ISBN: 978-1-003-47868-3 (ebk)

DOI: 10.1201/9781003478683

Typeset in Latin Modern Roman
by KnowledgeWorks Global Ltd.

*To my father's soul, my mother,
my husband and my children. You have made me
stronger, and more fulfilled than I could have ever
imagined. I love you to the moon and back.
To my sisters and brother,
my friends and everyone who encouraged me to make
all this possible.*

Contents

List of Figures

List of Tables

Preface

Industrial Internal of Things (IIoT) products and services collect a large amount of data including those related to users. This collected data are stored either locally on the smart devices or in the cloud. Although Internet of Things (IoT) expansions promise enormous benefits in productivity and efficiency, these devices often lack the security requirements we have become used to in the domain of desktops and server computing. Therefore, processing such large-scale IoT data can also lead to many security issues such as intrusion attacks, data leakage, user privacy, and traceability. The lack of robust security measures in defending IoT systems can compromise the IoT infrastructures and their data. Moreover, since IoT devices usually operate under tight resource constraints, IoT devices are a fruitful target for adversaries' exploitation.

This book presents substantial contributions to research on building a scalable, real-time IoT security framework. This book provides several novel strategies for identifying and authenticating devices, detecting network-known and zero-day attacks against IoT devices, and detecting malware executables targeting IoT and embedded systems. In this book, we consider three security aspects: IoT device identification and authentication, network traffic intrusion detection, and executable malware file detection.

The first key contribution is constructing a comprehensive survey that focuses on all security aspects and challenges facing IoT systems, including outsourcing techniques for partial computations on edge or cloud while presenting case studies to map security challenges and requirements in real IoT case scenarios.

The second main contribution is proposing a security framework named Behavioral Network Traffic Identification and Novelty Anomaly Detection for the IoT Infrastructures (BIN-IoT) that enables rules to constrain IoT device communications as per their given privileges. Based on our proposed novel IoT network traffic fingerprinting solution, BIN-IoT can passively select essential features from a sequence of network packets to construct a legitimate profile for each IoT device's network communications. This profile can help in identifying and authenticating devices within infrastructures. Moreover, any significant network traffic deviation can indicate an attack or a compromised IoT device within an IoT infrastructure.

An efficient near real-time IoT device identification and authorization system named Behavioral IoT Network Traffic Identification (BI-IoT) as one of the building blocks of the BIN-IoT framework is developed. BI-IoT combines the previously mentioned fingerprinting solution with Machine Learning (ML) techniques to authenticate devices connecting to a network. The proposed approach can automatically identify white-listed device types and individual device instances connected to a network. The proposed system improves the average device prediction F1-score up to 90.3%, which is a 9.3% increase compared with the state-of-the-art technique. Moreover, individual device instances sharing the same model and vendor as well as unknown devices are correctly identified with minimal performance overhead.

The third major contribution is developing another building block for the BIN-IoT framework, a near real-time IoT network traffic anomaly detection system named Behavioral Novelty Detection for IoT Network Traffic (BND-IoT) . The BND-IoT system's goal is to detect compromised IoT devices and malicious traffic within IoT infrastructures in real-time using novelty detection algorithms. The BND-IoT anomaly detection system can detect anomalous traffic from unseen attacks and malware traffic when the network model is trained with behavioral features extracted from the normal traffic only. The aim is to detect known attack patterns and zero-day attacks with High Detection Rates (DRs) and low False-Positive Rates (FPRs).

The fourth significant contribution is proposing a near real-time malware detection solution tailored for embedded systems, named DeepWare . It identifies malware by examining the binary file's executable Operation Codes (Op-Codes) sequence representations. We use Bidirectional Encoder Representations from Transformers (BERT) embedding, the state-of-the-art Natural Language Processing (NLP) method, to extract contextual information within an executable file's OpCode sequence. The BERT-generated sentence embedding is fed into a hybrid multi-head CNN-BiLSTM-LocAtt Deep Learning (DL) model. The hybrid CNN-BiLSTM-LocAtt model combines the advantages of the Convolutional Neural Network (CNN) and Bidirectional Long Short-Term Memory (BiLSTM) with the benefits of the Local Attention mechanism (LocAtt) to detect malware. DeepWare extracts the semantic and contextual features and captures long-term dependencies between OpCode sequences, improving the detection performance.

Authors

Salma Abdalla Hamad is currently a Senior Technology Security Compliance Manager at a telecommunication firm. She received a Ph.D. degree in Computer Science from Macquarie University, Sydney, Australia in 2021. She worked under the supervision of Professor Michael Sheng and Dr. Wei Emma Zhang. Salma obtained her bachelor's and master's degrees in Electronics and Communication Engineering from the Arab Academy for Science, Technology, and Maritime Transport, Egypt with distinctions in 2005 and 2009, respectively. Salma also possesses more than 19 years of working experience in both the government sector and the financial sector, where she executed several projects pertinent to Information Security. Her research interests are concentrated in the domain of Information Security, more specifically, for the Internet of Things, Smart Cities, and Smart Homes.

Quan Z. Sheng is a full Professor and Head of the Department of Computing at Macquarie University, Sydney, Australia. His research interests include the Internet of Things, service-oriented computing, distributed computing, Internet computing, and pervasive computing. Professor Sheng holds a Ph.D. degree in Computer Science from the University of New South Wales (UNSW) and did his post-doc as a research scientist at CSIRO ICT Centre.

Professor Sheng is the recipient of the AMiner Most Influential Scholar in IoT Award in 2019, the ARC Future Fellowship in 2014, the Chris Wallace Award for Outstanding Research Contribution in 2012, and a Microsoft fellowship in 2003.

Wei Emma Zhang is currently a Lecturer in the School of Computer Science, at the University of Adelaide. Her research interests include the Internet of Things, text mining, data mining, and knowledge base. She received a Ph.D. degree in Computer Science from the University of Adelaide in 2017. She has authored and co-authored more than 50 papers. She has also served on various conference committees and international journals in different roles such as track chair, proceeding chair, PC member, and reviewer. She is a member of the IEEE and ACM.

1 Introduction

The usage of Internet-enabled devices over the past two decades is becoming widespread and the Internet of Things (IoT) has become a massively distributed infrastructure comparable to or exceeding the Internet. The physical and digital worlds meet in IoT, in which the IoT is creating a seamless integration of physical things into communication networks, thus providing intelligent services for human life improvement [142]. Physical things connect to communication networks with wired/wireless technologies such as Ethernet, Bluetooth, Wi-Fi, and Zigbee, and improve their functionality with embedded sensors and , actuators. IoT devices impact and uplift the quality of various domains in both domestic and industrial sectors. IoT devices expand in a wide range of applications ranging from as simple as smart lamps to advanced intelligent manufacturing processes. There are various recent successful IoT applications in the market including smart homes, smart transportation systems, Industrial Internet of Things (IIoT), smart health care, smart grids, and smart cities.

Recent forecasting [262] reports that the amount of installed IoT devices worldwide is projected to be 30.9 billion units by 2025. The use of Radio Frequency Identification (RFID) in supply chain management and the advancement in different IoT-enabled technologies as illustrated in Figure 1.1 have significantly affected IoT growth in terms of the number of connected smart devices. The electronics manufacturing technology progress and innovativeness have significantly reduced the cost of IoT devices. Soon, everything will be embedded with small devices that connect to the Internet [237] to enhance various domains in our daily life, creating big data. Cloud computing has emerged to handle such big data from different types of devices in an IoT environment. Such information is distributed in different private cloud servers so that access is allowed only to legitimate users. Cloud computing as mentioned in [15, 52, 177] provides users with powerful computational ability and resources. Users with mobile devices or limited power devices can outsource some of the computations as well as data to be stored in the cloud.

Promising as it is, the cloud brings security challenges when users delegate sensitive operations to the cloud. While IoT and cloud technologies provide enormous benefits, they continuously face severe cyber threats that create various vulnerabilities. The heterogeneity of devices, services, communication protocols, data formats, and data generated from various smart devices in the IoT environment is one of the biggest concerns [11]. The smarter the network developments, the more challenges there are to secure IoT systems and maintain their privacy. To achieve new smart world requirements, systems need models for monitoring physical data and environmental conditions. All

1

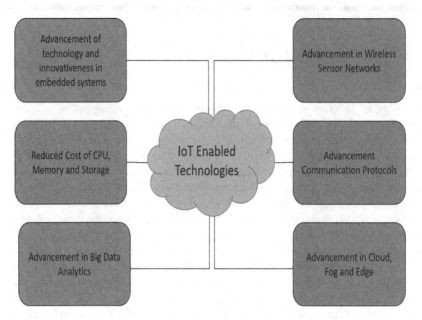

Figure 1.1: Technologies That Have Driven IoT Growth.

environmental data are collected by sensor nodes and then transferred to a remote location through a gateway or cloud.

Unfortunately, information risk increases with the increase in the deployment of smart things. IoT devices have different and varied natures, which leads to different types of security threats. These devices are vulnerable to different security attacks due to their constrained resources and the lack of protection methods [106, 237]. Promoted by the advances in IoT and cloud computing technologies for facing the mentioned security challenges, one of our primary goals is to develop a security attacks-free identification, and access control system which can be used for different IoT cases with a distributed cloud environment.

1.1   PROBLEM OVERVIEW AND MOTIVATIONS

In the context of IoT, two devices that communicate in an extensive network composed of multiple devices are not necessarily known to each other. As a result, additional strain is imposed on network administrators to manage the networks in terms of security, network resources, and device troubleshooting. Since these two devices initially do not trust each other and most of these devices have low computational abilities, no one can consider applying directly symmetric security solutions to create secure communications between them. Hence, the authentication of the source and confidentiality of data is required, which can be achieved using famous Public-Key Infrastructure (PKI) techniques. However, most of the public key encryption algorithms (e.g., Rivest-Shamir-Adleman (RSA) [245]) require intensive calculations, bandwidth, and complicated identity management systems for the devices to authenticate each other. Accordingly, these techniques are not suitable for low-bandwidth and resource-constrained environments. The challenge is to propose solutions that avoid the burden of PKI-based solutions while satisfying the security requirements. In such a case, the identification of these devices is crucial to the network administrators. Device identification is now an indispensable component in modern network management tools. It provides essential information such as the vendor, type, and specific communication features for a particular device.

Moreover, vulnerabilities within the IoT infrastructures arise from their physical devices security weaknesses, such as sensors and actuators, and flaws in their network components and communication protocols. Such defects emerge from the various challenges that exist during the implementation of the IoT devices, such as resource constraints (limited Central Processing Unit (CPU), Memory, Storage, Power), limited security patches provided by vendors, fast deployments of IoT devices and solutions, lack of standard IoT security best practices, and the heterogeneous nature of IoT devices and their related protocols [106, 188, 294].

IoT devices are simple devices, mostly built with minimal security concerns and low computational capabilities, thus coupled with various device-level vulnerabilities. Consequently, IoT devices offer an attractive target for adversaries. The common reasons for their security exploitation are [106, 188]:

- Their connectivity to the Internet.
- Their heterogeneous nature and ability to be remotely controlled.
- Usually, they are set with default configurations and passwords.
- Their lack of inclusive security mechanisms.
- Their ability to handle potential privacy and safety-critical data.

IoT vulnerabilities can supplant a range of cybersecurity threats targeting the IoT devices, the services that access the IoT device, and the IoT data itself. Recently, several real-world cyber attacks such as BASHLITE, Hajime, and Mirai malware [261, 283] have targeted the security weaknesses in the IoT infrastructures and affected their operations [58, 82, 106]. Researchers confirmed that the malware mentioned above compromised IoT devices and used them as pivots for launching Distributed Denial of Service (DDoS) attacks [82, 106].

The powerful Mirai malware has infected millions of IoT devices by brute-forcing the Telnet ports and obtaining unauthorized device access [126]. Similarly, adversaries strived to use BrickerBot malware [222] to realize access to vulnerable IoT devices. Brickerbot malware corrupted IoT device storage, leading to permanent damage to the compromised devices [222]. Moreover, Hajime, a successor of Mirai malware [13, 58], borrowed many tricks from

Mirai and other malware families to launch attacks on IP cameras, CCTV, routers, and other IoT devices. Furthermore, recently, researchers discovered a new Linux malware targeting IoT systems named Kaiji [33]. Kaiji is developed with a rarely used programming language, called Golang, and uses SSH brute-force techniques to attack IoT devices.

The ultimate goal of any security system is to preserve privacy as well as prevent attackers from compromising devices and exploiting vulnerabilities on any device in the network. IT general security should incorporate Availability, Authentication, Integrity, and Confidentiality requirements. In order to cope with the aforementioned IoT challenges and achieve the security goals, the below requirements should be integrated along with the general IT security requirements into IoT infrastructures:

- The first step in protecting an IoT infrastructure from attacks is in preventing unauthorized access by implementing a device identification and access control solution. Thus, identify any device that tries to connect to the infrastructure in real-time.
- Restricting the privileges of each of the connected devices, as per pre-defined rules. These pre-defined rules define where devices should be sited, by separating the devices into different privilege zones according to their communication and data importance.
- The IoT infrastructure also requires real-time solutions to detect attacks such as comparing the behavior of the current device, with a baseline generated during the normal communication of this device. To help in detecting rogue (misclassified as authorized) or compromised devices, thus limiting their network resources and privileges for further monitoring.
- An anti-malware solution is also required in the IoT infrastructures to detect malware binary files targeting IoT devices in near real-time.

All the requirements mentioned earlier are necessitated to maintain data security and privacy in IoT systems. Figure 1.2 represents an example of the security requirements and the challenges faced by all participating parties in the smart cities solutions. IoT solutions should integrate different techniques such as Machine Learning (ML) techniques in parallel with cryptographic techniques to provide the necessary services, as well as handle the challenge of an IoT device's limited computational power. Moreover, recently, malware and intrusions have been developed using more modern and innovative techniques to change their internal architecture to bypass detection. Strategies that have been used in the past few years to detect attacks, in general, might face obstacles in seeing IoT new forms of malware. Furthermore, malware changes its signature frequently. Thus, signature-based techniques by themselves are insufficient for addressing the intense threat facing IoT infrastructures.

ML and Deep Learning (DL) are subsets of Artificial Intelligence (AI) that decide a predictive model without being explicitly programmed to do so. Both

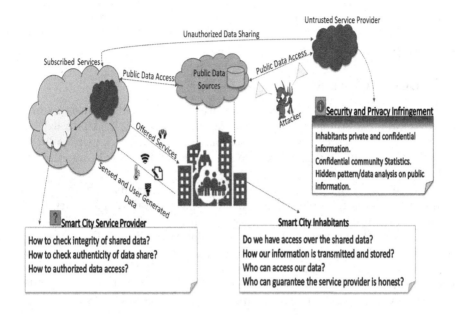

Figure 1.2: Case Study of Smart City Security Requirements and Challenges from the Perspective of All Participating Parties.

ML and DL techniques have the ability to process big data with complex characteristics efficiently. Hence are suitable for various applications in different domains, such as data classification, data prediction, image recognition, and Natural Language Processing (NLP). Accordingly, ML and DL can be used as building blocks in establishing a scalable, dynamic, and flexible IoT security framework that can detect both binary and network threats. In addition to this, such a framework should actively identify connecting devices to the network and detect any compromised IoT devices present within a network.

1.2 RESEARCH OBJECTIVES

This research aims to develop a security framework for IoT devices by employing an active approach to identify IoT devices connecting to a network and detect anomalous traffic and malware programs. The focus is on leveraging recent advancements in ML, DL, and networking standards to boost IoT security. With this in mind, the objectives of this book are divided into three phases: Phase 1 concentrates on building a real-time solution for identifying IoT devices connecting to a network using their network traffic traces and providing them with sufficient access privileges. Phase 2 focuses on developing a robust framework for detecting IoT anomalous network traffic. Phase 3 involves developing an anti-malware solution for detecting malware targeting embedded devices.

1.3 BOOK OUTLINE

This introductory chapter summarizes security threats to IoT infrastructures, context, problem statement, and motivation for this work. Afterward, an overview of this research's core contributions is presented, along with a broad outline of the remaining chapters. The remainder of this book is organized as follows:

Chapter 2 - Background and State-of-the-Art Security Solutions. This chapter covers background information and a state-of-the-art security techniques review. It overviews and justifies various authors' academic efforts and approaches in the cybersecurity field, including cryptographic techniques to secure IoT devices and methods of attack detection and analysis, and it compares them concerning the security service provided. This chapter also discusses the flaws and challenges in the current IoT security solutions and tools. Moreover, Chapter 2 expresses the requirements to secure IoT systems and the challenges developers and administrators face in managing and securing IoT systems and detecting intrusions and malware.

Chapter 3 - IoT Device Identification via Network-Flow Based Fingerprinting and Learning. This chapter gives a detailed description of our proposed security model design (BIN-IoT) and the . IoT device identification solution, namely BI-IoT. Firstly, the chapter presents the device identification problem and recent research targeting this issue. We then discuss the device's network traffic fingerprinting while giving the details of our novel fingerprinting solution. Then the integration of the fingerprinting solution with the ML networks is presented. Finally, we discuss the results and observations on the testing of BI-IoT on an IoT network traffic dataset. We compare our results with the other state-of-the-art solutions.

Chapter 4 - Behavioral Novelty Detection for IoT Network Traffic. In this chapter, we describe the implementation of our approach for IoT network traffic novelty detection system, named BND-IoT. BND-IoT is an IoT device anomaly detection system based on a neural network trained with the novel selected behavioral features extracted from benign traffic only. The chapter presents a brief description of the intrusion detection and anomaly detection problem in IoT networks, introduces the novelty detection techniques, especially the IF network, and discusses the importance of feature selection and extraction. Then an overview of the related work on network anomaly detection is discussed. We later present both the threat model and the proposed approach to detect compromised IoT devices connected to an IoT infrastructure. We then offer BND-IoT system architecture design and components. Subsequently, we report the experiments, set-up environment and experimental studies, and the results. Finally, a discussion of the proposed solution and observed challenges is presented.

Chapter 5 - Detection of Malware Targeting Embedded Devices. This chapter outlines our novel solution for the embedded devices' malware detection named DeepWare and a comprehensive explanation of its components. Firstly, the chapter gives an overview of the binary malware problem in embedded devices in general. It then discusses recent related research and its limitations. Then the chapter describes DeepWare; the new approach developed as part of the current study and, as such, is a contribution to the field. It then discusses the results and observations on the testing of DeepWare on newly collected datasets for the three commonly used CPU architectures in IoT devices. The chapter illustrates testing attributes used in the development of DeepWare. Moreover, it presents a comparison between DeepWare results and other state-of-the-art IoT static malware detection solutions.

Chapter 6 - Conclusion and Future Directions. This chapter concludes the research. It summarizes the findings and quantifications for completing the defined objectives of the research. Finally, it discusses directions for future research.

2 Background and the State-of-the-Art Security Solutions

The smarter the network develops, the more challenges need to be overcome to ensure IoT security and privacy [254]. To realize the recent smart world requirements, systems need models for monitoring physical data and environmental conditions. These models capture environmental data using sensors and then transfer the data to be processed or monitored through a gateway to the cloud.

This chapter presents the fundamental concepts, background information, and terms associated with this work. Besides, existing research work on the topic of IoT security is discussed to recognize the gaps and the extent to which this research work is proposed. In this chapter, we investigate and survey more than 100 information security-related works over the period of 2013-2020, while a number of them are specifically for securing the IoT with a focus on their diversity. We provide extensive discussions on securing cloud-based IoT solutions. The main focus of these discussions is on securing the information in transit between IoT devices and IoT applications, where most of the data processing and modeling tasks take place. These discussions include all security aspects and challenges facing the data in transit. Specifically, a number of common attacks that target IoT solutions are first discussed, while presenting the main challenges of IoT security (e.g., the resource limitation of IoT devices, which hinders the ability of such devices to do expensive computations for securing the data). Then we present a walk-through of some of the important security concepts and security requirements for cloud-based IoT while exploring possible solutions to mitigate cloud-based IoT threats. The chapter also discusses recent solutions for IoT cloud data sharing, including solutions for data access control, data, and user privacy, the integrity of IoT data, and outsourcing of computations for IoT devices to edge or cloud, respectively. Finally, we discuss threat detection including network attacks and executable malware attack detection approaches. The main content of this chapter has been published in a survey article [106].

2.1 THE INTERNET OF THINGS (IoT)

Since the term was first coined in 1999, the IoT has gained significant momentum in connecting physical objects to the Internet and facilitating machine-to-human and machine-to-machine communications [125, 303]. By offering the

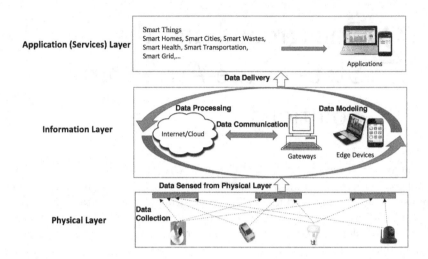

Figure 2.1: A Generic Architecture of IoT Systems. The physical layer collects data to be processed and modeled in the information layer, which then delivers the data to the application layer.

capability to connect and integrate both digital and physical entities, IoT has become an important paradigm that enables a whole new class of applications and services. In the near future, everything will be embedded with small devices that connect them to the Internet [237], to enhance various application domains in our daily life, such as smart cities, smart homes, smart transportation, smart health, and smart surveillance systems. Thus, IoT provides intelligent services for human life improvement [142]. IoT devices are heterogeneous in nature, leading to different types of security threats. Information risk increases with the deployment of a growing number of smart things. The limited resources owned by these devices make them vulnerable to security attacks, such as denial of service attacks [237]. These attacks can be performed on different layers of an IoT architecture.

2.1.1 IoT ARCHITECTURE

The generic architecture of IoT systems consists of three main layers as illustrated in Figure 2.1, including the *physical layer* (collection of data), the *information layer* (data analytics, sharing and storage) and the *application layer* (use of data).

The physical layer acquires data from sensors and network technologies making up an IoT ecosystem. In IoT services, sensors communicate with each other or with cloud-based servers. IoT devices have limited capacity to transmit large messages, due to the low radio frequency operation mode and low computational abilities. Data collection is done at this layer using actual physical devices, such as Radio Frequency Identification (RFID) sensors and others. Securing the physical device itself is as important as securing data transferred to or from the device. The limited hardware and application protection of IoT systems caused the most common threats in the physical layer. The devices at this level need to be secured, to protect the collected data and owners' privacy.

The information layer processes the collected data and then takes action according to the application's needs and requirements. The data are then either stored or sent for distribution to the application layer. This layer has attracted the attention of researchers in recent years, as it facilitates the development and integration of services [18]. The security and privacy components are required to ensure the collected data's authenticity and to maintain the user's privacy.

Different algorithms and techniques can be used in this layer for data processing, decision-making, and ensuring the protection and maintaining the privacy of the moving data, data owners, and data consumers. The details of threats and current solutions to secure and maintain data privacy at this level will be discussed throughout this chapter.

The application layer focuses on the feature specification for providing services, according to the final service implementation. This layer provides the user with the system's intended functionalities. It also inherits all the functionalities from the information layer [18].

2.1.2 IoT COMMUNICATION MODELS

IoT solutions allow people and things to connect anytime, anywhere, and with anything. This section provides some background on how things connect and communicate in terms of their technical communication models. Most of these models are supported and tested by the standard bodies. One of the main standard bodies that support the development and testing of IoT standards and communication models is the European Telecommunications Standards Institute (ETSI) [79]. IoT devices communicate with each other using three different communication models [231].

- Device-to-Device. IoT devices communicate directly with one another using any type of network such as Bluetooth and Zigbee.
- Device-to-Cloud. IoT devices connect directly to an Internet cloud service to transfer or store data. This type of communication usually uses Ethernet, Wi-Fi connections, or cellular technology.
- Device-to-Gateway. IoT devices communicate with a local gateway device such as a smartphone to access a cloud service.

2.1.3 IoT PROTOCOLS

There are a number of protocols and standards that help to empower IoT devices and applications. The most common IoT infrastructure and transport protocols used are IPv6 over Low-Power Wireless Personal Area Networks (6LowPan), Internet Protocol v4/v6 (IPv4/IPv6), Routing Protocol for Low-Power and Lossy Networks (RPLs), Bluetooth, Long Range Wide Area Network (LoRaWan), Zonal Intercommunication Global-standard (Zigbee) and Z-Wave. Regarding IoT data protocols, the most commonly used protocols are Message Queuing Telemetry Transport (MQTT), Constrained Application Protocol (CoAP), Advanced Message Queuing Protocol (AMQP), WebSocket, and Node [184].

2.2 IoT SECURITY

This section attempts to include all security challenges facing IoT solutions especially the information layer. We first discuss the current risks and limitations facing IoT devices. Then we study the common security attacks that can target IoT systems while understanding the significant security requirements to counter such attacks.

2.2.1 RISKS AND LIMITATIONS

This section presents some of the security risks and limitations for IoT devices in general across different layers, with more focus on threats targeting the Information Layer.

Physical Layer . To improve the security of IoT products, IoT devices need to be frequently patched and updated [27]. However, IoT devices have limited battery power as well as limited computational ability. In general, devices can be implemented on a broad spectrum of hardware and software. On the one hand, servers, desktops, and some types of smartphones are usually on the high end of the spectrum. On the other hand, embedded systems, RFID and sensor networks are on the lower end of the spectrum [169].

In regards to the hardware, IoT devices rely on microcontrollers which can vary in performance attributes. The most common available microcontrollers are 8-bit, 16-bit, and 32-bit microcontrollers [169]. As indicated in [169], there are significant sales of 4-bit microcontrollers for certain ultra-low-cost applications. These ultra-low-cost microcontrollers usually contain only a small number of simple instructions. Accordingly, a huge number of cycles will be needed to execute traditional cryptographic algorithms, thus making them time and energy-efficient for applications involving these devices [169]. Moreover, some microcontrollers have a very limited amount of Random-Access Memory (RAM) and Read-Only Memory (ROM). For instance, for TI COP912C [175, 268], the amount of memory can be as little as 16 bytes of RAM. Furthermore, at the bottom of the spectrum lies the RFID tags that are not battery-powered, which are powered by limited surrounding environmental power and have a limited number of gates available to serve the security requirements. A study on the constraints of such devices for cryptographic applications was performed in Saarinen and Engels [235].

Regarding the IoT software level, there are a number of operating systems that are designed to perform within constrained memory, size, and power used by current IoT devices, such as ARM Mbed [14], Brillo (Google Android Things) [269], Ubunto core [42], RIOT OS [286] and Contiki OS [61]. IoT OS security is very important and should support security services and privacy. However, most of these common operating systems are incapable of addressing the needed security requirements for IoT infrastructures [132, 168]. The challenge is to build a less vulnerable standardized, secure operating system for constrained devices that can provide all of the security and privacy services. Protecting IoT devices with the given limitations is a challenge. Nevertheless, security patching is considered one of these challenges that will expose IoT systems to a number of security risks.

Information Layer. There are a number of protocols that are used to manage IoT data and traffic as explained in Section 2.1.3. Unfortunately, these protocols mostly are deployed insecurely, resulting in sensitive data leakage, such as device details, credentials, and network configuration information [80]. For instance, it was reported [80] that the flawed implementations of MQTT which serves as one of the backbones of IoT and industrial IoT communications, expose sensitive information on the devices/servers to attackers using raw commands. Moreover, since MQTT can also be used for IoT devices software and firmware updates, this makes IoT devices more vulnerable to attacks [159]. Deshmukh and Sonavane [70] discussed the available security protocols and presented them with respect to the IoT layers. They analyzed the risks associated with each of the layers showing protection gaps. For instance, from this analysis, they observed that there is no fragmentation attack protection in the physical layer, network layer, transport layer, and application layer. They also highlighted that replay protection is not supported in the physical layer, network layer, and 6LoWPAN layer [70].

IoT-generated data from sensor devices flow to a cloud, Internet, gateway, or another device within the information layer. This data will be used in IoT applications. The information layer includes real-time and mobile data that feeds different IoT applications through the cloud, gateways, or edge devices and is sent from resource-constrained sensors. Accordingly, IoT systems have to deal with multiple threats or attacks which are described in Section 2.2.2.

Application Layer. The exposure and communication of IoT systems to the Internet have introduced new security requirements that applications should follow [190, 273]. Malicious modules can be deployed on nodes by attackers. Moreover, malicious users can attack vulnerabilities in operating systems (e.g., exploiting a buffer overflow vulnerability). Accordingly, software applications should run isolated from any other application, and other applications should not be able to intercept or alter their run-time state [112]. Furthermore, vulnerabilities in Web applications and IoT software can lead to compromised systems. For instance, Web applications can be exploited to either steal user credentials or inject malicious software.

2.2.2 IoT THREATS

Besides the particular malware that targets IoT and embedded devices, IoT infrastructures are at risk from all the attacks that threaten computer systems, network appliances, and cloud solutions. Table 2.1 summarizes these threats. These threats are mostly due to weaknesses in the overall architecture of the IoT solution. IoT malware can convert the IoT devices to bots that can be used to launch other network attacks such as DDoS.

2.3 CONCEPTS AND CASE STUDIES

With the increasing number of reported attacks facing IoT devices [107, 222, 22], researchers and industry are working on providing scalable and efficient solutions to detect and prevent IoT infrastructures from getting infected by malicious network traffic. This section presents the security requirements for IoT infrastructures while presenting case studies to map the security challenges and requirements in real IoT scenarios. The significant security requirements for all cases are data secrecy protection, integrity, access control, privacy protection, and the need for outsourcing computations to tackle the resource-constrained devices problem.

There are three main security requirements that need to be satisfied to protect any system, namely , *Confidentiality (Secrecy)*, *Integrity*, and *Availability* (CIA) [123]. Confidentiality ensures that the interpretation of a message is impossible for anyone except the targeted recipients. Integrity maintains the authenticity of the data in the IoT system. Availability ensures that the system can be accessed anytime and can serve in hostile conditions. Cyber-attacks pose severe threats to computer networking and violate the aforementioned CIA security requirements.

Table 2.1
Security Threats Facing IoT and Possible Countermeasures. These presented attacks include network, software, and hardware attacks that can target IoT solutions.

Threat-Attack	Description and Possible Countermeasures
Fabricating (forgery)	This attack fabricates a message and sends it to another person while pretending to be a certain person (e.g., pretend to be your manager and send a promotional email to HR for yourself). To counter this attack, the receiver of the message should authenticate the sender source. This can be achieved by Public Key Infrastructure (PKI) or similar signature techniques such as Attribute-Based Signature (ABS).

Threat-Attack	Description and Possible Countermeasures
Masquerade	Malicious attackers duplicate valid entities to be able to either access systems or impersonate a person. PKI can counter this problem by issuing a trusted identity certificate from a trusted Certificate Authority, while systems should always check on the validity dates of certificates, the certificate trust path (hierarchy) as well as revocation list.
Interception	Attacks data and user privacy. Encryption techniques, such as AES, should be used to counter it.
Data Alteration and Modification	The integrity of exchanged or saved data can be broken by modifying or deleting part or all of its content. PKI can be used as a way to detect and prevent data alteration. Hash functions as well as HMAC techniques can also help counter such attacks.
Illicit Access	Curious attackers try to access systems that they should not be able to access. To combat such attacks, access control techniques should be used. Access control techniques are either cryptographic, such as Identity-Based Encryption (IBE) and Attribute-Based Encryption (ABE), or non-cryptographic techniques such as Role-Based Access Control Technology (RBAC).
Replay Attack	In this attack, attackers resend correct data, which they gathered maliciously. Their target is to gain access to systems. There are a number of solutions to counter such attacks, such as adding timestamps to the messages.
Man-In-The-Middle Attack (M-I-T-M)	In an M-I-T-M attack, a malicious entity secretly eavesdrops on the communication and can alter the communication between two entities, while the two entities think that they have direct communication. Cryptographic solutions from encryption techniques, as well as mutual authentication techniques, are the common ways to prevent this type of attack.
Impersonation/ Sybil Attack	Sybil is a type of impersonation or fabrication attack. To prevent such attacks, PKI solutions with trusted identity certificates should be used as well as other signature techniques such as a digital signature or ABS.
Collision Attack	One or more illicit entities can collide together and combine their credentials to access data, that each one of them separately can't access. Each access control system has different ways to prevent such attacks. For example, the use of identity certificates using PKI can mitigate these attacks.
Timing Attack	It is delaying time-sensitive information, which can significantly affect time-critical applications. Such attacks can be prevented, by adding timestamps to data as well as by appending digital signatures.
Denial of Service (DOS)	This attack targets the data or system availability, in which malicious entities try to disallow system users from using the service/data. Different techniques can be used to reduce the effect of such attacks. For example, clustering (duplicate or triplicate) of all important servers or services.
Malware	Also called software attacks. Software attacks are the major source of security vulnerabilities in any system. These attacks have different forms such as worms, viruses, Trojan horses, and logic bombs. These attacks also can exploit a buffer overflow or inject malicious code into the system using different techniques such as Structured Query Language (SQL) injection. There are some famous malware attacks targeting IoT operating systems (OS) such as Mirai and the recently discovered malware , Silex [43]. To protect devices against such attacks, embedded security as well as physical security mechanisms should be implemented.

Threat-Attack	Description and Possible Countermeasures
Ransomware	Is a subset of malware that locks the data on a victim's computer, typically by encrypting it. Backing up data and ensuring business continuity by having disaster recovery plans is one of the usually recommended techniques to protect the data against this attack.
Side Channel Attacks	This is a non-invasive type of attack that is based on "side channel information" that can be retrieved from the encryption device that is neither the plain text to be encrypted nor the ciphertext resulting from the encryption process [20]. Side channel attackers can use different techniques such as timing attacks, hardware glitching attacks, and power analysis. There are a number of countermeasures for these attacks, such as blinding. However, the known countermeasures can work in certain scenarios but not for all. Accordingly, security should be embedded as one of the building blocks starting from the design phase.
Hardware Semi-Invasive and Invasive Attacks	Attacks on the device hardware such as decapping packages and using infrared emission analysis of the backside to find a location for attack, then using a laser to flip bits and break encryption. Other examples of hardware invasive attacks are micro-probing and modifying chips with Focused Ion Beam (FIB). Physical security for devices can limit such attacks [260]; devices should have physical safeguards against tampering or at least limit access to the hardware by putting the devices in a restricted place or secured with the appropriate locks or other tools.

Gathering information about a system breaches the system's confidentiality and, interrupting proper operations, affects its availability, and integrity [183]. For example, a Denial of Service (DoS) attack degrades the network resources affecting services availability. Simultaneously, malware alters the execution flow of an application, violating data and systems integrity [183].

Besides these main security principles, there are a number of other security services, that are specifically important for IoT systems. Table 2.2 describes the services that are needed in IoT systems. These services are required for IoT systems to mitigate security challenges and risks as well as other challenges due to the heterogeneous nature and resource limitation problem of IoT sensors.

The typical entities of an IoT system include a data sender (sensors), a data receiver (users/ sensors or actuators), and a Cloud Service Provider (CSP), as shown in Figure 2.2.

Table 2.2
Major Security Requirements for IoT Systems.

Requirement	Description Services
Confidentiality/Secrecy	It prohibits the interpretation of sensory data (in transit or at rest) from attackers. Allowing only legitimate users to access data.
Integrity	It is needed to ensure the detection of any modification of the data sent by sensors.

Requirement	Description Services
Authentication	It is needed for most secure IoT communications, to be able to identify the communication peer. This helps to prevent fabrications or impersonation attacks.
Privacy Protection	RFID tags are widely spread. Recently, it became, easy to track or identify objects using their tags, thus, raising privacy concerns. In addition, as wearable and implanted health devices increase their pace, our bodies will be connected to the Internet from small embedded devices. Consequently, people's personal information such as health care records or location must be secured and prevent unauthorized disclosure.
Conditional Anonymity	Protecting the anonymity of users in some IoT-related applications is crucial. Some users are reluctant to share data with others, to protect their privacy. Systems that don't protect users' anonymity may put the users at risk of being attacked (impersonation, tracking) [116]. However, most IoT applications need to authenticate the sender or receiver and ensure only particular authorities can trace this user (conditional anonymity) which can be used for emergencies.
Access Control (Authorization)	It is needed to limit and control access to data as per the pre-defined access rules/privileges.
Non-repudiation	Ensures that the data owner (source) can not the denial the previous data upload.
Availability	All sensors must be accessible while being used. The systems should be functional and immune against attacks. High availability and clustering solutions can help maintain service availability
Scalability and Interoperability	The number of users and devices communicating grows widely with the growth in technologies. IoT devices interact with different patterns, with a large number of entities. Capabilities-based access control mechanisms assure and complement IoT ecosystems security [161]. The proposed security protocols should be scalable enough, to handle a massive number of users/sensors. The systems should allow the integration and communication of devices from different environments.
Resilience to Attacks	IoT devices are usually inexpensive with limited physical protection. Sensors can also be in remote locations with limited monitoring, so they can be easily moved. Thus, the sensed data output can be modified, without anyone noticing. As a result, single points of failure should be avoided as well as different security measures should be enforced to secure systems against different attacks.
Attack and Threat Detection	Intrusion detection and anti-malware solutions are essential in the cybersecurity field for delivering a solid line of defense against cyber-adversaries.
Forward and Backward Security	Backward secrecy is to ensure that new group joining entities, can't decipher data created before they joined the group. Forward secrecy is to ensure that someone previously already had a key and after revoking them, they do not have access to future keys for the group.

The sensors send the sensed data to a cloud to be stored for later user access. Most of the receivers use more cloud services for sharing and securing data. An actuator does not send data but receives them to be used or further processed as per the application. The user is the entity that retrieves the specific data and accesses the shared data.

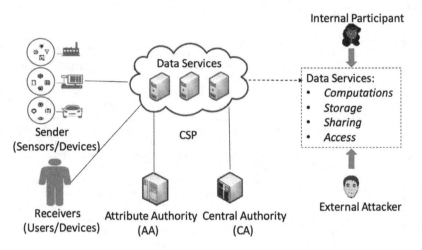

Figure 2.2: System Model for Cloud Based IoT Services.

IoT solutions can use cloud and edge computing not only for storing data but also for delegating computations to it. This can ensure data availability as well as provide the needed processing power for the security operation of the constrained devices. The cloud can participate in infrastructure management tasks such as keys and/or attributes issuing and revocation to facilitate the use of security services by IoT devices. IoT sensors can delegate partial computations while encrypting or decrypting data to a powerful computational device.

The delegation of computations can be divided into two models. The first model is to delegate the computations directly to the cloud. The cloud has the highest computational power that can assist in all types of operations. However, the latency in communication will increase by sending and receiving directly to the cloud. The other approach is to use edge devices as a delegator. Edge computing is a way to enhance cloud computing by performing data processing at the network edge (e.g., mobile phone, access point, laptop) near the source of the data. In edge computing, substantial computing and storage resources (cloudlets, or fog nodes) are usually adjacent to sensors [239]. Edge devices can be used by IoT sensors/devices as a bridge or gateway to the cloud. IoT devices can borrow some computational power from edge devices to do partial encryption/decryption during uploading/downloading data to the cloud. Thus, latency will be reduced while maintaining security and countering the resource limitation of devices. Moreover, the use of cloudlet computing can increase the scalability and availability (in case the cloud server is not responding or in case of network failure) of IoT systems. Outsourcing techniques to edge devices or clouds will be discussed in Section 2.5.

To ensure security, a third party such as an Attribute Authority (AA) and/or Central Authority (CA) may be involved. An AA is a trusted key authority, that checks a user's identity to generate attribute keys. Moreover, it manages the revocation or updating of user's related keys when needed. In general, an AA verifies the users' identities through a CA.

In an IoT ecosystem, most connected devices do not previously know each other. Accordingly, symmetric security techniques for securing communication will not be an effective solution. Thus, devices in such a situation usually depend on asymmetric security techniques and use a trusted third party for securing communications. However, some IoT devices are constrained in resources which might affect the key sizes and cryptographic operations. Accordingly, IoT device identification can work in parallel with asymmetric cryptographic techniques to identify and authenticate devices with an infrastructure.

The security and privacy issues in each of the three layers of the IoT system architecture have been studied independently under the themes of IoT device security, cloud/edge security, and application security. However, such independently developed security solutions miss the point that IoT services and applications are delivered by collecting data from the physical layer, processing them at the devices/edges/cloud, and then being accessed or consumed by users via applications and services. In general, there are a number of challenges facing the current techniques used to secure communication and data sharing in an IoT ecosystem. A brief discussion of the current security techniques' limitations is presented and clarified in the following.

Firstly, *authentication of source* and *confidentiality of data* are crucial security requirements that can be achieved by public key encryption techniques. However, most of the public key encryption algorithms use intensive calculations (e.g., RSA). Moreover, these techniques are mainly based on certificates for identifying and authenticating entities. The verification and management of certificates consume a large amount of computation and bandwidth. Since most IoT devices are based on ultra-low-cost microcontrollers, the excursion of such traditional cryptographic techniques will not be practical enough for IoT applications [169]. Therefore, a lightweight encryption algorithm with limited communication overhead is needed for securing communication between devices with limited resources.

Additionally, some of the current personalized authentication solutions may leak information which can cause *privacy* concerns [305]. In the IoT scenario, it is important to keep IoT users and devices anonymous from malicious entities as well as the communicating parties, except in emergencies and critical situations.

Moreover, there is a strong need for *delegating the expensive computations* to a powerful device to adapt the intensive computational encryption algorithms to the IoT systems with limited resources. There exist many techniques in the literature for outsourcing computations. However, the challenge is retaining the privacy and secrecy of the data and users, while using outsourced and lightweight IoT systems.

Furthermore, an essential need to integrate *adaptive and real-time Network Anomaly Detection* (NADS) and anti-malware systems with IoT infrastructures can discriminate between legitimate and suspicious network traffic and binary files. The NADS systems should be able to detect IoT-specific network traffic anomalies. At the same time, the anti-malware solution should detect malware targeting IoT systems with all its CPU architectures.

Finally, there is a demand for a security solution that can be lightweight enough to be used in resource-constrained IoT systems, while satisfying the required security requirements. To mitigate the aforementioned challenges, we set a number of objectives according to each of the following IoT case studies.

2.3.1 SMART GRIDS AND SMART METERS

A smart metering system is a type of IoT-enabled technology that supports high-frequency data collection compared to the existing metering systems. Smart grid users can securely manage and share their energy usage data. The captured data is analyzed and a report is created. This gives users the ability to inspect their energy consumption and correlate it with others (e.g., from the same suburban area). Accessing, analyzing, and responding to accurate and detailed data features, are crucial for an efficient use of the energy [59]. However, the more frequently the data is collected, the more the consumer's privacy is at risk, as it may expose the consumer's daily habits.

Smart grids and smart meters are essential for efficient energy management [59]. Ijaz et al. [120] showed the importance of not revealing the identity of users in smart grids. Moreover, a good number of surveys and technical papers discussed the security needs of smart grids. For example Martinez et al. [166] analyzed the privacy concerns in smart cities, including identity privacy, query privacy, location privacy, footprint privacy, and owner privacy [166]. Then, they proposed a model for mitigating these issues to ensure privacy in smart cities.

Due to the openness of smart cities and smart grid technologies, IoT systems are deployed in a vulnerable environment under the risk of a number of security threats. Regarding the smart grid, there are some security objectives for a practical solution, including:

*   *Data Authenticity.* In smart grids, if malicious users alter or forge energy usage data, this would provide misleading reports [281]. While the concern can be addressed by using cryptographic and integrity methods, e.g., message authentication code (Integrity) or digital signatures (Authenticity), other issues may be difficult to handle such as anonymity and efficiency. In other words, systems need to make sure data is from an authentic source (from a valid member) and that data content does not change.
*   *Anonymity.* Energy usage data contains customers' private information [76, 295]. Hence, safeguarding the identity of consumers is

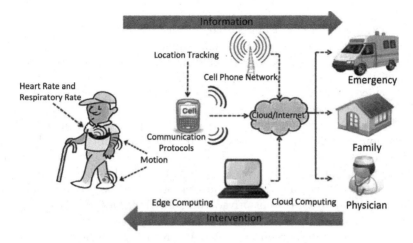

Figure 2.3: Smart Personal Healthcare Architecture.

crucial for such applications [73], to avoid raising privacy concerns and reluctance from the consumers to share their data.

- *Efficiency.* Data sharing in smart grids could be as big as sharing data of smart grids for a whole country [4]. Thus, the reduction of computation and communication costs is very important to prevent energy waste.
- *Availability.* The ability to access services or systems, even in hostile conditions, is critical [281].
- *Access Control.* Access to systems or data is restricted to authorized users only [241].
- *Confidentiality (Privacy).* Data should be protected at all times (at rest and while moving) and can only be interpreted by authorized users and devices [4, 241].

2.3.2 CLOUD BASED IoT HEALTH DEVICES AND PATIENTS RECORDS SHARING

A Personal Healthcare Device (PHD) nowadays is one of the fundamental elements of healthcare applications. The demands for healthcare for chronic and cardiovascular patients have increased significantly over the past years [142]. PHDs are movable medical sensors used by healthcare practitioners, to measure, record, and share user's biomedical information, as shown in Figure 2.3. The importance of PHDs is evolving, due to the increased demand of people to carefully monitor their health. Accordingly, such devices must be able to securely and easily share information with healthcare servers. However, the heterogeneity of these devices makes them difficult to manage and maintain and, thus are most vulnerable to attacks. For instance, in 2017 some implanted medical devices were attacked. WannaCry ransomware attacked devices running Windows OS and encrypted the data on the hard drive, making the devices inaccessible by users [258]. Such devices (wearable or implanted health devices) need to sense and send data to authorized recipients. Accordingly, fine-grained access control of patient data sourced from e-health devices is needed, while controlling who can access all data and who can access the data without knowing the identity of the patient (anonymity) as well as preventing access to user data if not authorized (e.g., not a primary physician or secondary physician). Moreover, a survey on medical device cybersecurity risks [205] showed that healthcare security professionals are concerned about medical device security, patient privacy, and data breaches, specifically as the patient's physiological data are transmitted through the Internet. Therefore, users need to be securely authenticated before using any of the IoT-based medical care services [142]. There are a number of security objectives for an efficient healthcare solution, including:

- *Data Authenticity (Integrity and Source Authentication).* Authentication is needed to certify that health data sources are credible and legitimate [62, 96].
- *Mutual Authentication.* Both sender and receiver need to ensure each other's authenticity [142, 311].
- *Conditional Anonymity of Sender.* The true identity of a patient can be revealed to, e.g., primary physician or police but cannot be revealed to, e.g., research students or any other secondary doctor [23].
- *Access Control.* Fine-grained access controls are required to define who can access the data [156].
- *Privacy.* Data should be protected at all times (at rest and while moving) [62, 96].
- *PHD and Health Implanted Devices Limited Computational Abilities.* It is critical to be able to offload encryption (partial) and decryption (partial) computations to proximity edge devices such as mobile phones or to the cloud, while preserving the data confidentiality and users' keys privacy [142].

2.3.3 SMART TRANSPORTATION (SMART CITY APPLICATIONS)

The main goals of smart transportation are to build a real-time intelligent public transportation system to reduce traffic congestion and increase safety and efficient energy consumption, to name a few [74, 162]. Sharing information among vehicles (mobile sensing) can provide location awareness, geo-distribution, and communication efficiency [282]. Such data can be shared from the vehicle's self-installed module or using an edge device (e.g., mobile

Figure 2.4: Smart Transportation Architecture.

phone) to the cloud as shown in Figure 2.4. However, security and privacy in the aforesaid infrastructure are challenging. Most cloud and fog service providers cannot be fully trusted. This will lead to the unwillingness of vehicle owners to share their collected data with strangers [282]. A trusted entity is needed, to prevent privacy violations, and promote cooperation in uploading vehicle data to fog nodes. Practical smart transportation systems should ensure that the following security services are in place:

- *Confidentiality* . The confidentiality of reports from vehicular IoT devices is one of the primary objectives to achieve [203].
- *Authentication of Source and Data Integrity.* Authentication is needed to certify that sources of vehicular sensing reports are credible and legitimate [137]. Blacklist-based authentication can help in preventing impersonation and Sybil attacks.
- *Anonymity and Privacy.* Privacy is a major concern, as the sensed data includes some information related to the drivers or passengers [203], such as their current location as well as their daily habits while driving. Accordingly, a system that can conceal the identity of the data source, while ensuring their authenticity is needed.
- *IoT Low Computational Ability (Secure Delegation to Cloud).* It is critical to be able to outsource partial encryption and/or partial decryption of the vehicular data to a fog device or cloud while maintaining the data and the users' privacy [144].

- *Non-repudiation.* The data owner (source) would not be able to deny uploading the previous data [144].

There are also several other security requirements that are needed for all three IoT cases:

- *Unforgeability.* An attacker should not be able to pretend to be an honest sender in creating an authentic signature text that can be accepted by the decryption algorithm. Forging keys and attributes or certificates should also not be allowed [4].
- *Collision Resistance.* It should be infeasible for two or more users to collide and combine their credentials to access data that each one of them separately cannot access [109, 204].
- *Unlinkability.* Attackers should not be able to link the used pseudonym (hidden name) with the true identity of the sender (multi-show) [112], e.g., given two messages and their signatures, no one should be able to tell if the same signer signed both messages.
- *Revocation.* User revocation is of great importance to IoT systems. A user/device may have a limited subscription period and has expired, or the device has been attacked or stolen. Accordingly, the communicating party needs to find out whether a user/device is revoked [297]. Any revoked user/device should not be allowed to read the data or authenticate himself. Efficient revocation is very challenging and it is especially important for a large-scale network.
- *Attack and Threat Detection.* The smart infrastructures should monitor the network traffic and executables running within the network and detect any anomalies, malicious traffic, or malware targeting the smart environments.

2.4 SECURITY MECHANISMS

This section explores some of the current or possible solutions to mitigate or detect the previously discussed security threats. To assure data security and preserve the privacy of users in IoT solutions, different security mechanisms are needed for each of the IoT architecture layers.

2.4.1 THREAT MITIGATION CONTROLS

Protection of both the physical and application layers should be ensured by including both software and physical security measures to safeguard data security and the data owner's privacy. IoT operating systems require an end-to-end security approach that should address security issues during the design phase. The software that is running on embedded devices needs to be secured, and regularly updated, and the solutions should have the ability to limit access to embedded systems to a need-to-use basis. Moreover, the embedded systems should provide a way for network administrators to monitor connections to and from the embedded systems. Furthermore, the systems should have the ability to integrate with third-party security management systems.

The operating system should support some important security features, such as securing the memory of the nodes [199]. On the nodes, all modules must not interfere with each other [199] and the implemented software should include cryptographic solutions for authentication and hashing. Moreover, communications among nodes should be secured against sniffing attacks, especially the RFID systems [255]. Encryption and access control services are usually used to protect communication against such attacks, which will be explained in securing the information layer in detail. Furthermore, to prevent the detection of RFID tags, there are a number of solutions in which an RFID reader transmits pseudo-noise. This noise is balanced by the RFID tags, which hinder the detection by malicious readers [240]. The physical security of IoT devices is currently of high importance. Protecting the hardware level of IoT systems is a parallel issue, which is under investigation and research in itself such as in Emura et.al [72]. Additionally, access restrictions on such devices and securing the information they contain must be ensured. Discussions on the current and possible solutions for securing the information and preserving privacy-related data in the information layer are detailed in Section 2.5.

2.4.2 THREAT DETECTION CONTROLS

The large number of devices connected to the IoT infrastructure and the nodes' heterogeneity are the main reasons for the infrastructure's security issues. The security compromise of a single node may lead to the breach of the whole system [147]. To hinder security breaches, attack, and threat detection, controls should be integrated with the IoT infrastructures. Attack and threat detection controls should monitor and analyze the activities and traffic on the IoT devices or network to detect possible threats. Discussions on the current and possible intrusion and malware detection system are detailed in Section 2.6.

2.5 THREAT MITIGATION SECURITY SERVICES

A good number of proposals have been presented in the literature on secure systems. Some of these proposals can be studied to see whether they can be implemented in IoT scenarios to face IoT-specific security challenges. In this section, we identify and arrange the literature targeting each of these requirements, and we study the recent research advances to secure the data and the users in IoT. Selected proposals are therefore briefed using security methodologies classification with respect to the targeted security service in Table 2.3. A brief classification of the security services that we cover in this chapter and the methodologies of each security service is illustrated in Figure 2.5, followed by the discussion of outsourcing techniques of security services.

Table 2.3
The Classification of IoT Security Methods.

Threat	Security Services	Methodology	References
Illegal access	Access Control (AC) (Confidentiality)	Identity-Based (IB) Encryption	[89], [98], [141], [284]
		Attribute-Based (AB) Encryption	[25], [28], [87], [93], [115], [118], [186], [223], [236], [289], [298], [306], [307]
		AB Signcryption	[50], [78], [156], [208], [226], [225], [309]
Disclosure of the user's identity with illegal AC	Anonymous ACl	AB Encryption with Hidden Access Policy	[127], [227], [295], [304], [305]
Fabrication or impersonation	Authentication	Public Key Cryptography (PKI)	[12], [250]
		IB Cryptography	[98], [143]
		Fingerprint Identity	[176], [221], [256]
User privacy breach	Anonymous Authentication	Anonymous Credentials	[40],[48], [145]
		AB Signature	[99], [129], [158], [164], [165], [206]
		IB Signature	[100], [292]
		Group Signature	[49], [134]
		Ring signature and IB Ring Signatures	[46], [55], [63], [174], [230], [247], [299],
		Signcryption, IB Signcryption and AB Signcryption	[4], [50], [68], [78], [102], [151], [156], [208], [225], [226], [274], [309]
Data corruption, alteration and manipulation	Integrity	PKI	[12], [250]
		Hash and HMAC	[64], [171] , [219]
		Cryptographic and Auditing Techniques	[35], [280], [285]
		AB Signature	[99], [129], [158], [164], [165], [206], [249],
		Signcryption, IB Signcryption and AB Signcryption	[4], [50], [78], [156], [208], [225], [226], [274], [309]
		Blockchain	[86], [131], [150] ,[180]
Attack on availability	Trusted Computing /Outsourcing	Server-Aided Signature	[31], [121]
		Homomorphic Encryption and Interactive Proofs	[57], [88], [90], [91], [146]
		Constant Size Cipher-text	[25], [202], [300], [307], [306]
		Proxy-Re-Encryption (PRE)	[17], [34], [148], [149], [272], [295], [297]
		Delegation of Computations	[16], [51], [94], [155], [191], [296], [300]

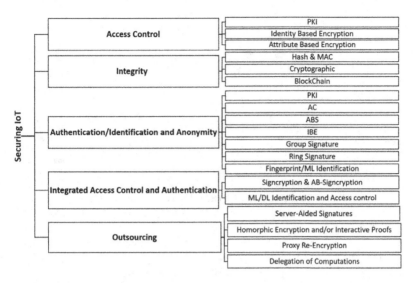

Figure 2.5: Classification of Security Services and the Current Methodologies.

2.5.1   ACCESS CONTROL

Any access control solution should manage who can do what on data. The system provides permissions and verifies the authorization of a user before accessing data. For IoT systems, access control solutions should take into consideration the special requirements needed by IoT devices, e.g., delegation support due to the limited computational power of sensors and devices.

Credential-Based Access Control (CBAC) techniques use users' credentials to access the requested data. Public key cryptosystems (PKC) use pairs of keys, i.e., a public key and a private key that belong to a certain owner. PKC can be used to accomplish two functions, namely *authentication* and *confidentiality*. Identity-Based Encryption (IBE) is a type of PKC that was first proposed in Boneh and Franklin [37]. In IBE, the user's public key can be a user's email address or any string that particularly identifies the user. A number of identity-based encryption schemes [89, 141, 284] and their variants were introduced and studied. Guo et al. [98] introduced identity-based encryption for lightweight devices, by eliminating the use of multiplicative group operations for encryption to reduce computations. Most of the IBE schemes are used for one-to-one encryption or at least need to know all the recipients' public keys; thus, they are not useful for large-scale data sharing.

Attribute-Based Encryption (ABE) was introduced in Sahai and Waters[236] to provide flexible, concrete authorization solutions. All users keep their authorization attributes and private keys. The data owner can

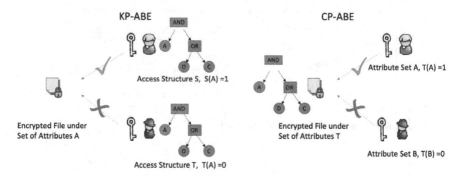

Figure 2.6: KP-ABE Vs. CP-ABE Illustration.

encrypt data according to certain access policies. Anyone who has an attribute set that fulfills the access policy can retrieve the data.

There are two variants of ABE as illustrated in Figure 2.6: Ciphertext-Policy ABE (CP-ABE), where ciphertexts are encrypted with access policies and keys include user's attributes [93], and Key-Policy ABE (KP-ABE) where keys are associated with access policies, and ciphertexts are associated with sets of attributes as defined in Goyal et al.[93].

The choice of which ABE variant should be used relies on specific applications. For instance, CP-ABE allows the data encryptor to decide who can access the data and choose an access policy, thus it is more suited for access control applications as compared to the KP-ABE schemes [202].

Many variants of ABE were proposed later. ABE with constant-size ciphertext was proposed [25, 306, 307], which produces less communication overhead. To limit the credibility of attribute authority, ABE that supports multiple authorities was proposed [223, 289].

To enhance or reduce ABE computations, online/off-line ABE was introduced [115]. The authors of [118, 265, 298] proposed solutions for efficient ABE revocation and leakage-resilient ABE, respectively.

ABE is required for rigid control of private data, such as Personal Health Records (PHR). Narayan et al. [186] proposed an attribute-based solution for PHR systems, which encrypts the patient's health records using CP-ABE that allows revocation instantly. However, the proposed solution allowed not only the patient's specific doctor but also practitioners to access the patient's medical record, without hiding the patient's identity.

An access policy needs to be protected, as it may include sensitive private information of a user. To prevent revealing the user's attributes, anonymous ABE has been discussed [304, 305]. In anonymous CP-ABE, the decryptor can not guess the ciphertext access policy (hidden access policy). In a number of anonymous CP-ABE techniques, the user has to perform a number of decryption trials to check if his attributes match the hidden access policy, which will lead to inefficiency in the system. To mitigate such a problem, Zhang et al. [304] constructed an anonymous CP-ABE scheme with match-then-decrypt, which improved the decryption efficiency. The shortcoming of this proposal is that it does not reinforce access policy updates [305]. Zhang et al. [305] proposed a solution that provides both user access policy updates and attribute privacy protection. However, the anonymity in this solution was only for the access policy. Moreover, the attribute and user revocation were not discussed in this paper.

2.5.2 INTEGRITY

The authenticity of data is needed, to ensure that the data content is not maliciously altered or deleted. Data integrity solutions should guarantee that a malicious user cannot make any change in the data (even if it is one bit) without being detected by the system as well as the receiver. In the literature, trap-door functions or hash algorithms, such as MD5 [219], SHA1, SHA256 [64] are functions that produce fingerprints of any size input data, while creating a fixed size output. This function reflects any single change in the input data, producing a totally different output fingerprint. Many other forms of integrity checking systems can be used, such as Hash-based Message Authentication Code (HMAC) [171]. HMAC is a Message Authentication Code (MAC) [233], combined with a one-way hash function and a secret cryptographic key. This allows the verification of both the authenticity and integrity of a message simultaneously. The strength of HMAC is based on the type of hash function and its output size as well as the size of the key. Usually, integrity-checking tools are combined with other cryptographic solutions to provide integrity and other services. For example, PKI and digital signature provide both authenticity of the sender and data integrity. Some techniques that will be discussed later in this survey, provide the integrity service in parallel with its targeted service, such as certificateless attribute-based signature (ABS) .

Data integrity can be ensured by distributing and replicating data over a set of nodes, which will reduce the possibility of a malicious entity imposing data on any node without being detected. Blockchain technology is a type of data replication disseminated among a huge number of nodes in diversified networks. Blockchain technology has appeared recently in the market, firstly used for the Bitcoin cryptocurrency [185]. Blockchain database provides balanced integrity while guaranteeing efficiency and stability [86]. There are a number of proposals in the literature [86, 131, 150, 180] that ensure data integrity and resilience using blockchain technology.

Blockchain can be used to provide a number of security services besides integrity because blockchain encrypts and hashes data using conventional cryptographic algorithms [178] and hash functions. However, the use of conventional cryptographic solutions adds processing overhead on the IoT devices which leads to slow transmissions [178]. Recently, a number of techniques have been proposed that use blockchain to provide security services with enhanced transmission rates. For instance, Roy et al. [97] considered blockchain as a provider of a complete secure IoT network. The authors introduced a blockchain-based framework that provides a number of security key elements such as decentralization, transparency, anonymity , and autonomy while considering the quality of service to ensure better transmission rates.

2.5.3 AUTHENTICATION

The ability of the broadspread IoT devices to collect huge amounts of information from their surroundings can cause privacy leakage issues and authentication problems [3, 313]. Moreover, some studies [9] have shown that the data collection can lead to the identification of individuals.

PKC is usually used to authenticate two entities with the use of an identity certificate [250]. It is used to identify and manage users and bind the user's public keys to their identity certificates, in a way that a third party can validate this binding. By binding PKC and trap-door functions (Hash), a digital signature is created, which adds the identity of the sender to any message [12]. This signature can then be verified using the sender's certificate. Identity-based (ID-based) cryptosystem, proposed by Shamir [250], eliminates the use of identity certificates and their verification, as the certificate management is time, bandwidth, and computation cost-consuming. ID-based cryptosystem uses any string that uniquely identifies a user as his/her public key. The string can be a passport number or an email address. Eliminating certificate validation reduces communication and computation costs, which makes the whole verification process more efficient. Many more implementations of IBE in the literature were studied. For example, Li et al. [143] designed a cloud architecture that uses ID-based cryptography for authenticating users and protecting data privacy.

The interaction of IoT devices and users produces personal information and leaves traces that can allow malicious users to trace the users' real identities [9]. For instance, when a user uses a mobile application to adjust house temperature, the system knows that this command comes from an authorized user. Accordingly, such systems cache the user's information that can reveal their actual identity [9]. Users authenticate themselves to any service before they can use it. The identity provider verifies the user to the requesting provider. This facilitates the authentication of identities. However, the involvement of the identity provider in all operations can give him the ability to trace users' connections to services [9].

To preserve a user's identity while achieving the needed authentication, the notion of anonymous authentication has been studied. Anonymous authentication can be achieved using a number of techniques according to the system's needs and computational capabilities. Ni et al. [194] presented a real-time steering system by utilizing vehicular crowd sensing while preserving the users' privacy. In the system, a Trust Authority (TA) issues anonymous credentials for each registered vehicle. A vehicle queries the nearby fog or edge while using a group signature. The information is gathered from the vehicles on roads, where the vehicles send real-time traffic information to the cloudlets while preserving their identities by using a signature induced from the anonymous credential. The requesting vehicle can then benefit from the recommendation path and follow the path to reach the target place. More importantly, the TA is able to identify misbehaving or forging data vehicles [195].

As IoT systems communicate with many heterogeneous devices, authentication between devices is needed to ensure they are communicating with the intended party. Cryptographic techniques are not the only available authentication methodology, especially for constrained systems. Such limited resource devices use small cryptographic keys for cryptographic operations. The use of small keys makes IoT devices vulnerable and easier to compromise. ML-based identification techniques are usually used to complement and work in parallel with the traditional cryptographic authentication techniques to compensate for the small key size problem and ensure higher identification/authentication security for such devices. Identification techniques usually require a unique fingerprint for each device or device type to be used by the ML algorithm for efficient automatic identification.

Several fingerprinting approaches are proposed in the literature, such as physical, wireless, and network trace fingerprinting [84, 135, 176, 221]. Fingerprinting IoT devices is usually a challenging task due to the high number of available device types and used protocols [29]. Most of the existing network traces fingerprinting approaches that are used to feed into identification solutions [176, 221, 256] mainly focus on flow-based features or only include information from a limited number of network layers and protocols. Due to such limitations and the heterogeneous nature of IoT devices, these approaches might not be applicable in real-time systems.

Authentication or identification usually includes identifying the device or user [107]. However, such identification makes users and devices traceable, hence, their privacy is threatened. There are a number of approaches that tackle this problem, such as Anonymous Credentials [40, 48], Attribute-Based Signatures [129, 158, 249], Identity-Based Signatures [100, 165, 292], Group Signatures [49, 134], and Ring Signatures [55, 63, 174, 230, 247, 299].

2.5.3.1 Anonymous Credentials

Anonymous Credentials (AC) was presented in Chaum [48], and then formalized in Camenisch and Lysyanskaya [40]. These schemes are one of the essential components in privacy-preserving identity management solutions. AC allows users to authenticate themselves to systems while confirming possession of credentials to service providers without being identified. In an anonymous credential system, a credential-issuing entity identifies users by pseudonym and issues a credential to them according to this pseudonym. This pseudonym is related to the actual identity of the user. A user can prove to a service provider that they own a pseudonym, which they received according to their credentials from an authentic organization [145]. Anonymous credential systems are also capable of providing fine-grained access control [62]. However, it is not efficient to support complex predicates when compared to ABS and ABE [145].

2.5.3.2 Attribute-Based Signature Scheme

Magi et al. [164] formulated Attribute-Based Signature (ABS) and grouped it into KP-ABS (Key Policy ABS) [249] and SP-ABS (Signature Policy ABS) [165]. Liu et al. [158] proposed an Anonymous ABS for anonymous authentication while outsourcing most of the users' computations to cloud servers, to be able to use this authentication system with low-power computational devices. In Kaaniche and Laurent [129] used ABS as AC while allowing the users to reveal the required information only to any service provider to ensure unlinkability as well as preserve the anonymity of the user. The authors relied on a non-interactive protocol in their derivations.

Guo et al. [99] proposed an attribute-based signature for electronic health records that ensured the integrity of data within the blockchain and disclosed only the related evidence or data as per the attributes used. Bricer and Kupcu [206] proposed a multi-authority ABS that uses only a subset of the attributes to reduce the computation and communication overhead.

2.5.3.3 Fuzzy Identity-based Signature

A fuzzy Identity-Based Signature (IBS) was presented in [100, 292], which enabled users to use part of their attributes to generate signatures. However, IBS does not protect the signer's identity.

2.5.3.4 Group Signature Scheme

A group signature scheme is an approach that allows any group member to anonymously sign data, and the signature will be seen as a group signature. It was originally introduced in Chaum and Van Heyst [49] as a concept. For example, a large company can create a group signature scheme for its employees. An employee from this group can sign a message, and it is acceptable to verify that an employee signed that message, but not who was the specific signer. Then Khader [134] proposed an attribute-based group signature scheme. It protects the signer's identity while proving only that the signer's attributes match the policy. The verifier can only identify that the signer is a group member. Attribute-based group signatures can provide conditional anonymity, which makes it more useful for attribute-based signatures in certain situations. Conditional anonymity is a feature that the group manager can identify the signer of any signature. The identity of the signer can then be revealed if required.

2.5.3.5 Ring Signature

Ring signatures were first proposed by Rivest et al. [230]. It is a type of adjustable signature that protects the signer's identity within a group. A user can be a member of a ring spontaneously, in which he can create a group of his choice. The other users in this group might not know that they are ring group members [145]. Messages can be signed anonymously by any member of this group. By verifying the message, it can be justified to the verifier that a ring member signed the message, without being able to trace the identity of the signer. Ring signatures could be used in applications that need signer anonymity, while not having the complicated group formation stage. A number of different techniques were discussed [55, 174, 247] since the first proposal and first introduction of ring signature [63, 230], respectively. The idea of ID-Ring signature was proposed in 2002 [299]. ID-Ring signatures can provide the same features of identity-based crypt-systems, without using the high computational bilinear pairing.

The Ring signatures and ID-Ring signatures can provide total anonymity. However, conditional anonymous authentication cannot be achieved as no one can tell who signed from the ring, even in emergencies.

Cloud infrastructure provides solutions that are reliable and ensure data availability at a low cost, which is useful for both service providers and consumers. The external storage of a user's data on a cloud server owned by third parties, and the ability to access this data from the Internet put data and users' privacy under concern. Accordingly, the security and privacy of users' data have become an active research question recently. To prevent cloud insider attacks, Bleikertz et al. [35] introduced a fine-grained privilege levels approach to provide users' privacy and integrity as well as the ability to perform cloud maintenance. Raykova et al. [227] proposed a technique that hides private data in policies from cloud insider attacks. The authors defined an access control system with two sides: the cloud side which has limited access to information, to be used by the cloud providers, and the client side, which depends on access control cryptosystems.

2.5.4 INTEGRATED ACCESS CONTROL AND AUTHENTICATION

A number of applications require data secrecy while ensuring the authenticity of the origin. For example, e-Health systems as well as Personal Health Records (PHRs) share a patient's personal health with a number of expected users, such as doctors and insurance providers. The e-Health and PHR service providers usually use the cloud for PHR data storage. Storing health data on a semi-trusted cloud raises security and privacy concerns. The main need for such systems is to guarantee that health data is available for legitimate authorized users.

Healthcare fraud or abuse as well as misdiagnosed patients can happen [96] unintentionally, or intentionally, if a malicious unauthorized user accesses the data and modifies the e-Health data before the doctor accesses it. This may lead to an incorrect prescribed treatment for a patient, which could cause a threat to the patient's life. To prevent the patient's identity, both the patient's privacy and the authentication of the target receiver should be achieved, during the process of uploading e-Health information to the cloud. Securing the privacy and anonymity of patients while sharing PHR data in cloud computing environments is an evolving issue. Accordingly, fine-grained data access control solutions that ensure the confidentiality , authenticity , and anonymity of users are essential.

Signcryption provides confidentiality and authenticity simultaneously [309]. It executes both signature and encryption, with less computational overhead compared to Sign-then-Encrypt approaches. Attribute-Based Signcryption (ABSC) was proposed such as Chen et al. [50], combining the functionality of ABS and ABE. There are two types of ABSC: *signcryption-policy attribute-based signcryption* (SCP-ABSC) and *key-policy attribute-based signcryption* (KP-ABSC). On the one hand, there are two policies associated with the users' attributes in SCP-ABSC, and a key is tagged with receiver attributes and sender attributes. On the other hand, in KP-ABSC, everything related to policies and attributes is swapped. Emura et al. [78] described dynamic ABSC, that does not need to re-issue users' secret keys during the access structure updates. Then, a number of ABSC schemes [4, 50, 208] have been proposed in the literature. A number of proposals handled computational limitations of mobile devices [156, 225, 274].

It is noted that the computations and communication overhead of ciphertext policy ABSC [78, 156] boost linearly with additional attributes. This computation and communication boost has motivated the construction of a ciphertext policy ABSC scheme with a fixed size and computational cost. For preserving the users' privacy and identities, there is a need to use multi-authority-based attribute encryption to limit the capabilities of a single authority, while including a semi or partial ID verification for collision, without exposing the identity of either the sender or the receiver. ABC and ABSC can be used in many applications. But, it is still not widely spread on mobile and IoT applications due to its high computations.

2.5.5 OUTSOURCING OF COMPUTATIONS

Cloud computing can provide users with powerful computational ability and resources. Users with mobile devices or limited power devices can outsource their intensive computations or store their data on the cloud. Cloud brings security challenges when users delegate private operations on it, such as signature generation. The cloud can sign user's messages, without the knowledge of the user. To handle the semi-trusted or untrusted cloud or service provider problem, a server-aided signature scheme approach was proposed in Jakobsson and Wetzel [121]. The techniques still need expensive computations, which cannot be afforded by IoT systems. Another methodology centers around outsourcing with methods presented [57, 88, 90, 91, 146] that use homomorphic encryption or interactive proof systems. However, such methods were proved by Gentry [91] that they are not efficient enough for systems with limited computational power and resources, such as IoT devices.

In IoT, many devices are resource-constrained. Accordingly, a number of security solutions targeting limited computational power systems started to emerge [272]. The authors introduced a solution that considers limited capability devices and delegates the CP-ABE computations to powerful computational devices. The produced ciphered text is then either re-sent to the sending device or forwarded to the cloud service provider for storage or sharing [210]. In another work, Odelu et al. [202] proposed a CP-ABE technique that generates constant-size ciphertexts and private keys, which is more feasible for IoT devices.

Proxy Re-Encryption (PRE) was introduced in Blaze et al. [34] and formally studied in Ateniese et al. [17]. PRE allows a proxy that holds a re-encryption key created by the data owner, to re-encrypt ciphertexts (that are already encrypted with Alice's public key), to be encrypted with Bob's public key instead, in a way that only Bob's private key can decrypt the newly re-encrypted ciphertext. PRE and Attribute-Based Encryption (ABE) can be used for securing data access stored on the cloud.

PRE is an ultimate solution especially for cloud storage since we cannot fully trust CSP, while the useful property is the ability to do the ciphertext conversion without revealing the corresponding plaintexts. Integrating ABC and PRE can produce a flexible and practical fine-grained access control data-sharing solution. Ciphertext-policy Attribute-Based Proxy Re-Encryption (CP-ABPRE) technique was first introduced in Liang et al. [149]. In this technique, a proxy is able to convert a ciphered message ciphered under an access policy to a ciphered message ciphered under another access policy. [148] constructed the first CP-ABPRE that has only one trusted authority for generating keys. Thus, it is not feasible to be used in scalable big data systems. Massive data systems should be supported by multiple authorities to allow scalability and separate authority roles. In 2017, Yin and Zhang [295] proposed a CP-ABPRE with attribute privacy protection, by hiding the access policy, to protect a user's identity.

IoT devices are usually lightweight devices with very limited hardware resources. Unfortunately, the computational cost of cryptographic algorithms (e.g., pairing and exponentiations) are costly. IoT devices cannot afford such a cost. Accordingly, it would be desirable that such devices can delegate part of the extensive computations to a gateway or cloud, as illustrated in Figure 2.7.

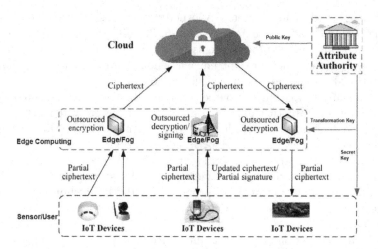

Figure 2.7: Delegating Partial Encryption and Decryption to the Edge or the Cloud.

In the literature, there are a number of conducted works that target delegating the computations of encryption, decryption, and/or revocation of ABC, as shown in Figure 2.8. Green et al. [94] described a CP-ABE solution that outsources the decryption process. Zhou et al. [312] described a technique that delegates both the encryption and decryption of CP-ABE to the cloud. Their proposed encryption process has two access structures (T1, T2) connected by an AND root node. This solution requires three exponentiations on the user side, which still needs plenty of resources. Asim and Ignatenko [16] presented a CP-ABE technique that outsources both the encryption and decryption extensive computations. During key generation, a user's secret key and a transformation key are created. This transformation key should be given to a proxy or cloud service provider. The proxy uses the transformation key for partial decryption of the ciphertext to ElGamal ciphertext [77], only when the user's attributes match the policy. Then the resource-constrained device

Figure 2.8: Classification of Research in Outsourcing According to What to Be Outsourced.

can decrypt the generated ElGamal ciphertext with limited computations. An efficient Outsourced ABSC (OABSC) was presented [51], which borrows computational resources from a third-party cloud to partially decrypt ciphertexts. In this process, users need to select and record their secret keys, which will allow them to do the final decryption of ciphertext. Usually, the ciphertext is stored on the cloud. When users request to access the data, they need to get verified by presenting their attributes. Then a transformation key created from the user's key should be sent to the cloud system to partially decrypt data. This will then return a partially decrypted ciphertext. Then the receiver can verify the correctness of the transformed ciphertext, before decrypting to its original form using their kept secret key.

Yang et al. [291] proposed a multi-cloud-based outsourcing for decryption while preserving the receiver's attributes from being disclosed. In Nguyen et al. [191], OEABE outsourcing scheme was introduced, which outsources ABE ciphertext-policy encryption. This proposed solution targets the delegation of the most expensive computations in the ABE encryption to a cloud while maintaining the confidentiality of data against both external and internal attackers. The encryption process requires only one exponentiation on resource-constrained devices. However, the solution does not target the revocation of users and attributes, or the delegation of attributes management (issuance or revocation) to the cloud or proxy. Zhang et al. [300] proposed an energy-efficient KP-ABE decryption outsourcing that takes into consideration of the cipher-text size, making sure that the cipher-text size is constant to reduce the communication overhead.

Regarding attribute revocation, a number of CP-ABE mechanisms that manage attribute revocation have been presented. Yu et al. [297] used a semi-trusted proxy for instant attribute revocation. In their proposal, the proxy transforms the ciphertext, as well as refreshes all the authorized users' secret keys. Yong et al. [296] described a CP-ABE technique that outsources both decryption and attribute revocation. This technique uses attribute versioning to accomplish attribute revocation. Liu et al. [155] presented a technique to outsource decryption as well as attribute revocation. Their technique concentrates on ciphertext updates as well as updating the user's keys, reflecting the revoked attributes.

There are a number of solutions that tackle the outsourcing of encryption, decryption, or attribute management (i.e., revocation) to the cloud. However, to the best of our knowledge, no technique nor framework in the literature integrates all the needed services for IoT devices. IoT frameworks need to provide conditional anonymous authentication, and fine-grained access control with receivers and sender privacy and data integrity. Such techniques need to outsource encryption, decryption, revocation management, and access policy management to a gateway or the cloud.

2.6 ATTACK AND THREAT DETECTION

Traditionally, malicious activities within an infrastructure are detected using signature-based techniques, which cannot detect unseen anomalous traffic or malware files. Recently, numerous research approaches have been introduced in the literature that use emerging techniques to detect malware and intrusions. Both intrusion and malware detection systems are used to monitor host, and network-based environments. A host-based Intrusion Detection System (IDS) and end-point malware detection monitor the activities happening on a computer system only [147]; thus, it requires an agent to be installed on the host. Since most IoT devices have limited resources, host-based attack detection solutions are not suitable. Remote attack detection solutions, such as Network Intrusion Detection System (NIDS) and cloud-based malware detection monitor IoT system network traffic and binary execution flow remotely to detect attacks [124]. The remote monitoring techniques do not degrade the performance(s) of IoT devices or the software running on the network. Moreover, these solutions can be installed over the network, cloud, or edge devices to aid any changes happening on the network topology, thus providing a more scalable solution.

As this work concentrates on the development of remote threat detection approaches, in this section, we overview remote attack and threat detection methodologies, reviewing different techniques for detecting malware within an IoT infrastructure.

2.6.1 MALWARE DETECTION

The threat of developing malware attacks is continuously expanding, which has provoked researchers to analyze and defend the systems against these new malware variants. The theory of malware detection is to trace and analyze executable files to establish malicious intent. Accordingly, malware analysis becomes an essential component in emerging attack detection solutions. Analysis of malware provides an understanding of malware behaviors, thus facilitating the building of proper detection and mitigation methodologies against the attacks [104]. Typically, anti-virus software uses a signature-based method to recognize malware. The instructions flow in the malware executable is captured to obtain a signature that uniquely identifies the malware from an extensive database of known malware signatures [217]. Malware analysts use a wide range of malware analysis techniques to detect malware effectively. The main malware detection techniques are based on static and dynamic analysis. Some reported methods to detect malware by blending some dynamic and static features [56, 157, 163]. However, hybrid analysis usually requires a large amount of time and resources to analyze the binary file's characteristics; thus it is not applicable for real-time scenarios [229]. The malware analysis variants extract unique features that can be then fed to Machine Learning (ML) and Deep Learning (DL) networks to detect malware.

2.6.1.1 Static-Based Detection

Static malware detection involves analyzing malware binary files without executing them. Static detection is performed by reverse-engineering the malicious code and examining its instructions. The analysis can capture static information about the binary file [104]. The static features can show low-level behaviors of the binary file structure itself, such as the Executable and Linkable Format (ELF) file header or capture high-level instructions within the binary such as strings, OpCodes, Control Flow Graph (CFG), Application Programming Interface (API) calls, and import functions, etc. [229, 266, 301]. Static analysis can detect malware quickly and effectively with a low false positive rate. However, static detection methods have disadvantages, such as most of their approaches cannot effectively detect malware that uses obfuscation and encryption techniques [196].

Operation Code (OpCode) is one of the commonly used static features for malware detection. In the assembly language, an OpCode is a single instruction command, such as CALL, ADD, or MOV executed by the processor. OpCode sequence is a high-level static malware detection feature extracted from the binary file by a disassembler [220, 228] that usually comprises the program's execution logic [53]. In this work, we focus on static malware detection using OpCode sequences and DL techniques.

2.6.1.2 Dynamic-Based Detection

The dynamic analysis [124, 213] usually executes the malicious binary in an isolated environment such as a Sandbox, a Virtual Machine (VM) or an emulator to monitor its behaviors. Running the actual malware binaries can resolve some of the limitations in the static detection methods such as code obfuscation and encryption [287]. However, dynamic analysis is usually resource-intensive and can be bypassed in various ways [217]. Since IoT devices have diverse microcontroller architectures (e.g., MISP, ARM, PowerPC, x86) [114] and are usually resource-constrained, it is difficult to correctly set up the required environment for running and monitoring the IoT executables.

Dynamic detection can involve monitoring the running processes, file system activity, memory, registry, and the network traffic generated during the malware execution. Network monitoring is one of the common dynamic malware analysis and detection techniques. A network traffic monitoring and malware detection solution usually captures traffic generated during the malware execution to understand the communication channel and protocols used by malware [128]. Network traffic monitoring to detect intrusions is commonly known as IDS. More details about IDS are discussed in Section 2.6.2.

2.6.2 INTRUSION DETECTION

An Intrusion Detection System (IDS) is a mechanism responsible for monitoring network environments to detect malicious activities. Many research publications and industry solutions were concerned with IDS for general computer and network systems. With the recent enhancement of embedded systems and the widespread use of IoT solutions, cybersecurity experts and researchers have been concerned about integrating IDS with the IoT architectures and devices to deal with cyberattacks [147]. There are four major security approaches to detect anomalous network traces, namely, signature-based, anomaly-based, policy-based, and hybrid.

2.6.2.1 Signature-Based Detection Techniques

This is also called misuse detection. This technique compares the network traces with a set of signatures that represent the patterns of known attacks. These known signatures are usually stored in signature databases that are frequently updated with the latest recorded attack patterns [69]. This technique cannot detect any new intrusion or zero-day vulnerability. A zero-day vulnerability is a software or firmware flow that is not known by the manufacturer or security professionals. A zero-day attack happens when attackers leverage this flow to compromise systems [200].

2.6.2.2 Anomaly-Based Detection Techniques

The network traffic anomaly detection objective is to detect any anomaly in network traffic. Anomalies are known as patterns in network traffic that have different behavior than normal traffic. Detecting these malicious patterns automatically can help network and system administrators detect compromised machines in the network as well as help them to analyze network behavior for forensic purposes [259]. The anomaly detection techniques usually apply ML algorithms to detect anomalies. The main types of anomaly detection techniques are:

- *Outlier Detection.* Most of the research proposals in anomaly detection typically use network traces from common types of known attacks to train a network model, thus these trained network models can identify and detect such anomalous traces. Network traffic can then be classified as normal or abnormal.
- *Novelty Detection.* This is a mechanism for detecting unusual behavior by training a model using normal behavior patterns only. Then the classifier is expected to recognize any behavioral changes or anomalies. It is a practical approach for monitoring network traffic, since we usually do not know what an unseen anomaly or a zero-day attack would look like.

Anomaly-based techniques in general can be used to detect unseen anomalies [251]. However, most of the current security approaches do not provide efficient methods to analyze traffic traces and confirm that it includes anomalous traces from a zero-day attack. Indeed, many researches and solutions aiming to fix this issue have been made known in the literature [81, 108, 113, 172, 179, 182, 187, 197, 234, 253, 288]. A major concern of this approach is that it cannot distinguish all legitimate behaviors of users.

Recently, there has been an increase in research focusing on using novelty detection techniques in different application domains, especially domains that include large datasets. It is difficult in critical and complex systems to understand the relationships between different layers and components of themselves and to define abnormalities. This makes novelty detection techniques offer a good solution, where models learn normal behavior from examples of positive instances. The novelty detection approaches are capable of detecting unseen malicious traffic and zero-day attacks. Accordingly, it can be a useful base for IoT network anomaly detection.

One-class classification is usually the approach used for novelty detection. One-class classification can be used to tackle the zero-day attack problem. This classifier aims to identify objects of a specific class (normal traffic). The classifier is trained using data only from the normal traffic class. During the classification, any object from any other class is identified as negative (anomaly). However, the absence of negative class training samples complicates the classification problem [207]. One-class classification can act as a baseline measurement for a number of applications such as anomaly detection [5, 39, 45, 160].

The one class novelty classifier $N(\gamma)$ with γ parameters is trained with the normal positive instances X. The model is expected to distinguish any other instance that varies from the normal patterns by calculating prediction scores $S(X)$ for test data X, where such data usually contains unseen data patterns. The larger the $S(X)$ scores, the larger the probability of having anomalous data in the test data. To set decision boundaries, a threshold T is defined, where $S(X) \leqslant T$ for the case of normal traffic and $S(X) > T$ for the case of anomalous traffic [215].

There are numerous traditional novelty detection techniques in the literature, which can be categorized into a number of categories, such as distance-based and density-based [215]. However, these traditional techniques usually can not handle big data and data with different density regions [66]. An ideal anomaly detection classifier should (i) be easy to tune, (ii) have fewer parameters to set and have fast run-time, (iii) can scale up to handle big data, and (iv) be simple and consistent with different types of data and different types of anomalies [270]. Isolation Forest (IF) [153] is one of the novelty detection approaches that can handle big data challenges, and it can be easily tuned and used with different types of data and anomalies. IF focuses on separating anomalous data, not on trained normal traffic. It usually consists of an ensemble of trees, where partial models and sub-sampling are used to isolate anomalies, with low computations and linear time in comparison to traditional distance and density-based novelty detection approaches [153]. The ability to use sub-sampling enhances the possibility of hosting a fast execution online anomaly detection solution that uses low memory to fit big data infrastructures [153].

2.6.2.3   Policy-Based Detection Techniques

Policy-based detection techniques capture the outliers by defining the allowed and blocked network traffic using a set of rules [47]. This technique can mitigate some of the previously mentioned concerns for both signature-based and anomaly-based techniques, such as the detection of unknown or zero-day attacks and the lowering of False Positive (FP) cases from legitimate unseen behaviors. However, the accuracy of the defined policy depends on the security specialist's experience who created the rules. Moreover, conflicts between inter-policies and intra-policies might happen, triggering more than one rule which can cause confusion as well as increase FP and f=False Negatives (FN) cases.

2.6.2.4   Hybrid Techniques

Hybrid solutions are simply the integration of different detection methods. This hybrid technique is receiving considerable attention recently [7, 209], because it can overcome the limitations of previously mentioned techniques. Henceforth, it can increase the Detection Rates (DRs) without increasing the False Positive Rates (FPR) while also gaining the positive aspects of both the signature-based and anomaly-based detection techniques [290]. Nevertheless, the hybrid technique still has tough challenges, especially in the architecture and classification decisions [7, 271].

As discussed above, each of the presented approaches has both strengths and weaknesses. The signature-based technique usually triggers fewer FP alarms but cannot detect attacks that do not have a known signature in its database. Similarly, the anomaly-based technique can be efficient enough to recognize novel attacks. However, it can trigger high FPRs, as it may be normal traffic with unknown patterns but is mistakenly identified as malicious traffic. Recent research in literature targeted introducing new mechanisms and systems to resist intrusions in standard network protocols. However, most traditional IDS techniques are incompatible with IoT devices' limitations or with the IoT complex network structures [264]. Accordingly, extensive research on IDS using ML methods is essential to secure IoT infrastructure and protect IoT privacy.

2.7   SUMMARY

This chapter presents background information about IoT and its components. In particular, it defines the IoT systems, their architecture, layers, and protocols. The chapter gives an understanding of security threats and security requirements of the IoT, as well as approaches to address different threats and security challenges. Since IoT systems should be developed with due consideration of security, privacy, and trust, security requirements should be addressed in all of the architecture layers of IoT systems. Furthermore, constraints introduced by the IoT, such as big data, and distributed and low-powered devices, should be considered during the solutions design phase. In this chapter, IoT security threats and possible countermeasures for each threat are investigated. Challenges to IoT devices and the exchange of data between IoT systems, edge devices, and the cloud are evaluated. Then a walk-through on some of the vital security concepts and security requirements for cloud-based IoT is presented. Three IoT applications are used as case studies to reflect the security solutions needed for cloud-based IoT systems. In addition, we offer a detailed survey of security mechanisms that address both threat mitigation and detection controls. In particular, we discuss different security services and related challenges in IoT, namely confidentiality, access control, authentication, privacy, integrity, resource-constrained IoT devices, and IoT-targeted attack detection. Several promising opportunities are emerging with the evolution of IoT and cloud computing for facing IoT security challenges. However, a complete solution has not yet been implemented. With IoT devices communicating over a hostile network, these devices are more likely to get exploited, which puts the overall data trustworthiness at risk. Therefore, IoT network traffic should be carefully monitored, IoT devices should be identified efficiently, and any malicious activity should be automatically detected. Moreover, IoT devices interact in different patterns with different entities. The number of users and devices communicating together within an IoT ecosystem is growing significantly with technology growth. Accordingly, research should focus on providing an efficient and scalable security framework to fit different types of IoT devices with different CPU architectures.

3 IoT Device Identification and Fingerprinting

Nowadays, an increasing number of intelligent devices and smart sensors are connected by IoT techniques and have helped people to manage and improve their lives. However, security issues are emerging in IoT, among which things identification is one of the challenges in that various solutions of different vendors, standards, protocols, and community groups coexist. Chapter 2 noted that current cryptographic technologies are not necessarily effective alone while authenticating IoT devices to a network. In this chapter, we address the challenge of IoT device identification by analyzing a sequence of packets from its high-level network traffic, i.e., network-flow data, and extract unique flow-based features to create a fingerprint for each device. We adopt supervised ML techniques for the identification task. The proposed approach can automatically identify white-listed device types and individual device instances connected to a network. Moreover, we propose a security system model design, named BIN-IoT that enables enforcement of rules for constraining the IoT device communications as per their given privileges. Unknown or suspicious devices with abnormal behaviour can be identified, and their communication is restricted for further monitoring as discussed in Chapter 4. We show that the presented approach is effective in identifying white-listed device types with average accuracy up to 90.3%, which is a 9.3% increase compared with the state-of-the-art technique. Individual device instances having the same model and vendor as well as unknown devices are correctly identified with minimal performance overhead. The main content of this chapter is from Hamad et al. [107].

3.1 OVERVIEW

IoT interconnects the physical world with the digital world with a significant impact on all life aspects. Accordingly, in recent years, IoT is overspreading in both industrial and household sectors. These heterogeneous smart devices need to be integrated into the common infrastructure. Despite the number of benefits in terms of usability and flexibility, these devices also bring security concerns [248]. Among these concerns are device identification, authentication, and privacy.

Recently, a large number of attacks that target the security and privacy of IoT systems were performed [139]. For instance, attacks that target faking or changing the readings of sensors or maliciously accessing devices, to capture sensitive data or to change the device into a state useful for the attacker [139]. Security of smart sensors usually, is one of the last, if not the least, priorities for IoT companies and developers, which result in a number of successfully exploited device vulnerabilities [22]. With this prompt increase in the number of heterogeneous IoT devices that join the network and the growth of the network complexity and size, appropriate solutions are needed to help network administrators manage the networks. A number of security concerns can be tackled by using well-designed identification and authentication mechanisms. This allows network administrators to manage and enforce security controls for individual devices. Automatic identification of devices, not only allows the administrators to ensure devices are connected to the correct zone with the right privileges, but it also gives them information such as product vendor, product type, operating system, and software version of a particular device. Such information can help them locate vulnerable devices, unauthorized/rogue devices, or devices that need patches or updates in the network [176].

IoT device identification is challenging due to the existence of various IoT device types, communication protocols, and control interfaces. An untrusted or unknown device can masquerade as another device from the network's white list and gain network access. Moreover, rogue or compromised devices might try to gather information or disrupt the normal functions of other devices within the infrastructure. This can be detected by comparing its behavior with its baseline fingerprint.

Traditionally, device identification and/or authentication depend only on cryptographic authentication techniques. However, IoT devices usually are constrained. Accordingly, these devices use small cryptographic keys to be used for cryptographic operations. The use of small keys makes IoT devices vulnerable and easier to be compromised. ML identification techniques are used to complement and work in parallel with the traditional cryptographic authentication techniques to compensate for the small key size problem and ensure higher identification/authentication security for such devices.

ML device identification techniques are recently introduced in the literature to automate the identification process. Each device should have a unique fingerprint to be used by the ML algorithms for efficient automatic identification. A number of methods for device fingerprinting are introduced in the literature. Such techniques include physical, wireless, and network trace fingerprinting. Physical fingerprinting is based on the device's physical features, such as device hardware Clock Skews [135]. Wireless fingerprinting techniques, focus on certain protocols to identify devices [1, 83]. Network traces fingerprinting, or extract features from the Internet Protocol (IP) network traces of the device in question, such [176, 221, 256]. Existing fingerprinting solutions that target device identifications, either focus on a limited number of protocols, which makes them not suitable for the heterogeneous nature of IoT devices, or on recorded low identification rates. Thus, they are not applicable to be used in real-time IoT networks (these solutions will be discussed in more detail in Section 3.2). Therefore, we present a novel fingerprinting technique that identifies device types and individual device instances with high identification rates.

This chapter focuses on identifying real-time IoT devices, that are connecting and communicating using IP-Network. We propose an approach that uses a passive behavioral fingerprinting technique to identify devices. The fingerprint is created by extracting features from a sequence, a set of packets ordered by time stamps. These features are extracted from the network packets header and payload to generate a unique fingerprint for each device. These behavioral and flow-based fingerprints can be used as a baseline to continuously monitor the device's behavior while connected to the network. The continuous network monitoring and anomaly detection solution (BND-IoT) are detailed in Chapter 4. To identify an unseen device connecting to the network, we collect a limited number of sequential packets to create the fingerprint. Then we apply multi-class classifiers on the extracted features to predict the device name. We extensively compare several widely-used classifiers and find Random Forest consistently shows the best performance. Details and analysis of the used features are discussed in Section 3.5.1.

Contributions. The key contributions of this work are as follows:

- We present a self-learning real-time security model design, named behavioral network traffic identification and anomaly detection for the IoT infrastructures (BIN-IoT). BIN-IoT can aid network administrators to control access to IoT devices by identifying connecting devices and provide each identified device with its predefined privileges. Each device has a predefined set of privileges including a communication zone. Any behavioral alteration for any of the communicating IoT devices from the behavioral baseline can be detected by a continuous random moving window fingerprinting of the devices. Hence, the identified compromised devices are quarantined for further monitoring.
- We propose a novel network fingerprinting technique for IoT infrastructure that captures the behavior patterns of Internet Protocol (IP)-enabled IoT devices network traffic. This behavior-based fingerprint is applied to the identification solution (BI-IoT) that uses ML algorithms to identify IoT devices connecting to a network. The presented fingerprint study shows that it can identify device types accurately with only a small sequence of packets.
- Our proposed device identification system (BI-IoT) can identify connecting device types as well as individual devices from the same type (same vendor and model), which is an essential feature to manage the network and the access control solutions for each device, not only the device type.

The rest of this chapter is organized as follows. Section 3.2 presents related work, reflecting the differences between previous solutions and this proposed approach. In Section 3.3, we present the threat model to identify IoT devices as well as to detect compromised IoT devices connected to an IoT infrastructure. This section also discusses the specific IoT infrastructure challenges that need to be managed by the proposed model. We then propose BIN-IoT, our security model design that securely identifies devices and then connects them to their communication zone. We then propose a device identification methodology in Section 3.5. Section 3.6 presents the experiments and results. Finally, Section 3.7 concludes the chapter.

3.2 RELATED WORK

In the era of the IoT, almost every device that we see or hold can be connected to the Internet, and there is a considerable amount of data that can contain personal or corporate data shared all over the Internet. Accordingly, authentication of the source and confidentiality of data are significant security requirements [106]. As it is a matter of IoT devices communicating over a hostile network, it is more likely for these devices to get exploited, which puts the overall data trustworthiness at risk. Therefore, IoT network traffic should be carefully monitored, and any malicious activity should be automatically detected.

IoT device identification and fingerprinting is a recent research topic and is in its early stages. It is evolving with the growth of the IoT industry. Franklin et al. [84] presented a passive fingerprinting technique that can identify different wireless driver implementations on network-connected devices. Abdelnur et al. and Francois et al. [1, 83] proposed two fingerprint approaches based on the protocols used, with high detection accuracy. These solutions analyze the behavior of certain well-known protocols, specifically, SIP as a protocol. Being limited to certain types of protocols might limit the scalability of the solution. Moreover, IoT heterogeneity results in a diverse number of protocols used within IoT devices. Hence, it is difficult to use such techniques in IoT environments. In contrast, we propose an approach that can be used with a variety of connection technologies. Our proposed approach is tested on a dataset with devices that communicates with more than four types of connectivity technologies.

A general purpose of device type identification was described by Radhakrishnan et al. [221]. The main aim of their solution is to recognize the device types connected to the IP network. Their solution is based on packets' inter-arrival times to extract features related to a particular application. Our work is influenced by their proposed solution. We execute a number of experiments to select the best features. The analysis of these experiments showed that including statistics of inter-arrival times between packets, enhanced the overall accuracy. Siby et al. [256] proposed IoTScanner, which identifies devices by visualizing Medium Access Control (MAC) layer traffic. This solution is more concerned with the presence of devices on the network. The proposed technique can be very useful for high-level network mapping. However, it will add a lot of processing loads if used periodically for identifying devices and comparing their fingerprint against the baseline. On the contrary, our approach uses only a limited number of sequential packets to fingerprint any device, with minimal performance overhead.

IoT Sentinel was presented by Miettinen et al. [176], which proposed a passive fingerprinting technique that analyzes the headers of packets for a connecting device. The authors used a Random Forest single classifier to identify each device independent from the other 26 devices. Their solution is capable of detecting unknown devices. The authors evaluated their solution with off-the-shelf 27 different IoT device types. The presented solution achieves an accuracy average rate of 81% for identifying IoT device types only. A shortcoming of this solution is that it is not capable of identifying individual device instances from the same device type (same vendor and model). Moreover, the prediction accuracy in this solution is lower than other state-of-the-art solutions. But, it has considered a diversity of IoT devices with different protocols. The prediction accuracy of 10 devices from the dataset was around 50% or below. Our approach complements this solution to achieve higher prediction rates as well as recognize different devices having the same vendor and model. We propose a device fingerprinting technique that captures the patterns in a limited number of sequential packets. Moreover, features from both packet headers and payload, are included, to create a unique fingerprint for each individual device, not only the device type.

3.3 THREAT MODEL

This section first discusses the research challenges around IoT infrastructures and then presents the threat model for IoT infrastructures. The section also gives a high-level overview of our proposed solution with the technical details in Section 3.4.

3.3.1 IoT INFRASTRUCTURE CHALLENGES

The nature of IoT infrastructures adds several challenges to the security systems [105]. The most prominent challenges are:

- **CH1**: Most IoT devices have limited resources. This makes using only cryptographic authentication unreliable as well as hinders the possibility of implementing host-based intrusion/anomaly detection.
- **CH2**: The manufacturers of IoT devices are increasing the frequency of producing new IoT devices. However, a significant number of these manufacturers do not provide any security fixes or updates or even pay no concern to IoT security, which increases the possibility of producing vulnerable devices [106]. Having such vulnerable devices

connected to the network increases the attack and data leakage risk for the overall infrastructure.

- **CH3**: Usually, IoT devices generate limited traffic patterns, which makes it challenging to generate a behaviour baseline that can be used for identifying devices as well as detecting any misbehaving device. The limited traffic may result in a decrease in the identification rates and an increase in anomaly detection false positives.

3.3.2 ADVERSARY MODEL

We consider an adversary model, where malicious attackers are attempting to gain access to the network assets by compromising vulnerable IoT devices in the infrastructure. Gaining access to the infrastructure assets can give attackers the ability to exfiltrate sensitive information or use the compromised devices as bots for further attacks. In this work, we are also considering already compromised IoT devices within the infrastructure that produce unusual network behaviors.

The primary goals of a security system are to preserve the privacy of IoT device data and effectively detect malicious traffic and protect infrastructure-connected devices from being exploited. Thus, the objective of this work is to provide a solution that can identify any device joining the infrastructure in real-time to provide it with sufficient access privilege and to uncover malicious traffic targeting or generated from the IoT devices. Identifying connecting devices gives the network administrators the privilege of taking appropriate defense or protection methods in case of infection, such as prevention of malicious traffic to communicate with inter-devices or isolating compromised devices.

In our work, we assume the IoT devices used for collecting the classifiers' training data are not compromised at the collection time. The collected traffic data for devices joining the network is used for training and testing the identification classifiers. In contrast, standard device communication traffic data is used to train the anomaly detection classifiers.

By considering the challenges mentioned above, we propose a real-time device identification and a self-learning anomaly detection system that learns the baselines of IoT devices using their sample traffic. The approach uses semi-supervised learning algorithms to identify labeled devices and detect anomalies from unlabeled network traffic generated from IoT devices connected to the network. In particular, this proposed solution can be applied as a network appliance or on an edge device within the industrial sector's infrastructures, to provide adequate access privileges to the IoT devices and to detect malicious traffic within the IoT infrastructure. The technical details can be found in the next section.

3.4 BIN-IoT SYSTEM

In this section, we present a security identification, authorization, and security enforcement model design. We propose BIN-IoT, a system for IoT device identification, security enforcement, and network-traffic anomaly detection. IoT individual device identification tackles CH1 which can manage and restrict access to any device joining the infrastructure. Moreover, in Chapter 4 we describe the network traffic anomaly detection module (BND-IoT) to mitigate the risks arising from CH2. To consider CH3, in this work, we capture a small network traffic sequence to identify the IoT device or detect anomalies in the network traffic.

The ultimate goal of any security system is to preserve privacy as well as prevent attackers from compromising devices and exploiting vulnerabilities on any device in the network. This can be achieved by

- Identifying any device that tries to connect to the infrastructure in near real-time.
- Restricting the privileges of each of the connected devices, as per pre-defined rules. These pre-defined rules define where devices should be sited, by separating the devices into different privilege zones according to their communication and data importance.
- Comparing the behavior of the current device, with a baseline generated during the normal communication of this device. To help in detecting rogue (misclassified as authorized) or compromised devices, thus limiting their network resources and privileges for further monitoring.

The proposed security enforcement model design consists of two main stages, as demonstrated in Figure 3.1. The first stage of the model is used when an unseen device tries to connect to the infrastructure. The model captures a sequence of the network traces for the device in question. It then identifies this unseen device after creating a fingerprint, using a trained classifier, that is trained with white-listed devices. The decision of the classifier can be either an unknown device or a white-listed device with identified device name and type. If the decision is unknown, the network connection between the unknown device and the internal network will be rejected. Otherwise, the identified device name and type will be sent to the zoning stage. This stage takes the authorization matrix for each device as a second input. The authorization matrix reflects the trust level of each device as per its known vulnerabilities and latest patches. The zoning stage then allows the device to connect to either the trusted network or to the restricted network as per the pre-defined authorization matrix. The trusted zone devices can communicate with all devices and services within the infrastructure. However, the restricted zone devices will communicate only with restricted devices in the same zone.

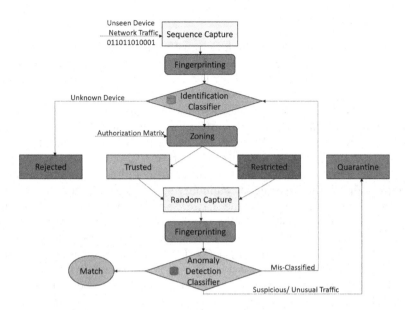

Figure 3.1: System Model for IoT Device Identification and Security Enforcement.

The second stage of the model is during the normal device communication within the network. A random moving window sequence of packets captured is taken from any device in the two zones. This captured traffic is then fingerprinted and applied to a trained classifier, to compare the device behavior with a stored baseline. The decision of the classifier can be either device misclassified or misbehaving. Misclassification can result from the device being wrongly identified as a certain device from the Stage 1 classifier, while it is a different device from the same white list. Accordingly, the new predicted device name and type will be sent to be re-zoned as per its related authorization matrix. Misbehaving of devices happens if a device is compromised or wrongly identified as a white-listed member. This can be detected if the fingerprint does not match any of the baselines in the database. Hence, the device network communication will be quarantined and monitored for further analysis. This stage can be integrated with a Security Event and Information Management (SEIM) solutions to take further actions and alerts.

3.5   DEVICE IDENTIFICATION METHODOLOGY

The proposed device identification approach (BI-IoT) has two components: fingerprinting and classification. The fingerprint features selection as well as the identification problem description are detailed here, leaving the selection of the best classifier to identify devices in Section 3.6.2.

3.5.1   FINGERPRINTING

To create a fingerprint for each device, we extract unique behavioral and flow-based features from the header and payload of (Ethernet, IP, and Transmission Control Protocol (TCP)/ User Datagram Protocol (UDP)) a sequence of packets. The packet sequence is ordered by time stamp. Feature selection is usually used to reduce feature space by selecting features that include informative unique information. This will increase the accuracy of the classifier and reduce computational cost. We focus mainly on packets originating from the device in question. But some features are extracted from bidirectional traffic as well. The source Media Access Control (MAC) address is used to identify the source of packets. We calculate summary statistics (minimum, maximum, first quartile, median, third quartile, mean, variance, and inter-quartile) and Fast Fourier Transforms (FFT) for a number of the used features [10, 193]. FFT can identify the periodicity patterns within network traffic flows.

Capturing long samples from the network packets of the device in question, usually seize more useful information. However, to achieve real-time performance, smaller length inputs need to be used, which can be insufficient to capture the unique information for each device. Accordingly, the accuracy of the classifier can be influenced. To tackle this conflict, we run a number of experiments to find the optimum trade-off between quality and real-time performance of the classifier as shown in Section 3.6.3. As a result of these experiments, we choose to extract features from a sequence of 20-21 packets.

The fingerprint for each device includes selected behavioral and flow-based features. Behavioral features , usually capture information that reflects the behavior of the device such as ports used and patterns [182]. Flow-based features include information flow, such as in IP destinations as well as statistical information, such as Inter-Arrival-Time (IAT) [182]. The fingerprint for each device network trace consists of a set of 67 features. It includes Ethernet packet size, IP packet, header sizes, and TCP payload size features. Some of these features are added as an influence [181]. In addition, we include TCP payload data offset. Data offsets give data start information to the upper network layers. We add it due to its significant influence on the uniqueness of the fingerprint during our experiments. Moreover, Time-To-Live (TTL) for TCP and UDP packets is used as a feature. TTL is the number of hops allowed for the packet, which is used to limit the lifespan of packets in any network. TTL feature for TCP is hinted at [173]. Furthermore, we include IAT summary statistics and the 10 highest magnitudes of FFT. IAT measures the delay between the following packets. It was previously used as a feature in [152, 221, 232]. Packet direction is added as an integer to represent the direction of each packet. A count of the number of IP destinations, source, and destination ports, within the selected sequence communication, are added as features as well. In addition, we include packet rate, which is the count of transmitted packets within a specific time period (5 secs). Finally, we use the TCP window size feature. TCP window size is the number of packets that can be sent and acknowledged in a single acknowledgment by the receiver. It reflects the device memory as well as processing speed and was suggested as a feature in Bezawada et al. [29].

3.5.2 IDENTIFICATION

We form the problem of device identification as a multi-class classification problem [29], because we are identifying different device types and different individual device instances. Given a set of authorized devices (white-listed), the classification task is to identify the class of the device in question.

A number of commonly used machine learning techniques in literature are examined to choose the best model for the selected features. We compare the classification performance using nine classifiers: Adaptive Boosting (ABOOST), Latent Dirichlet Allocation (LDA), K-Nearest Neighbors (KNN), Decision Tree (CART), Naïve Bayesian (NB), Support-Vector Machines (SVMs), Random Forest with 50 estimators (RFC50), Random Forest with 100 estimators (RFC100), and Gradient Boosting (GBOOST). The settings and results of these experiments are detailed in Section 3.6.2.

3.6 EXPERIMENTS

In this section, we present our experiments on the identification methodology, which is a key component of our proposed security enforcement model design. We use real-world IoT device data. We describe the settings and dataset, analyze different classifiers, and present the results and our analysis.

3.6.1 SETUP

We describe the experimental settings in this section, including dataset, evaluation metrics, and implementation.

Dataset. The dataset used in this chapter is the same dataset created and used by the authors of IoT Sentinel [176]. It consists of traffic traces from 31 off-the-shelf IoT devices saved as packet capture (pcap) files. Among these, there are 27 different device types. A number of pcap files for each device are recorded during the initial communication between the device and the infrastructure. The devices used in this dataset have more than 4 types of connection technology, such as Wi-Fi, Zigbee, Ethernet, and Z-Wave.

Metrics. To consider the effect of false positives and false negatives on the identification accuracy, we decided to use F1-Score [242] as the validation metric. It is the "weighted average of the precision and recall" [242]. Equation 3.1.

$$F_1 = 2.\frac{1}{(\frac{1}{recall}) + (\frac{1}{precision}))} = 2.\frac{precision.recall}{precision + recall} \qquad (3.1)$$

The general formula for f_α is shown in Equation 3.2 [276]. α can be chosen by the deploying organization. The lower the α, the stricter the network [173], as the classifier will value precision more than recall. Thus, fewer unauthorized IoT devices will be identified as white-listed devices. However, this also will increase the False Alarm Rate. Moreover, organizations can set a prediction decision threshold, tr, to set the network strictness.

$$F_\alpha - Score = (1 + \alpha^2) * \frac{precision.recall}{(\alpha^2.precision) + recall} \qquad (3.2)$$

Our goal is to classify the network stream for every device trying to connect to the network, as originating from an authorized white-listed device or unknown device. Accordingly, we apply the classifier to the extracted features. The classifier produces a vector of posterior probabilities of classification P^s for the test input. We then examine these calculated posterior probabilities. If any of the posterior probabilities P_i^s for device d_i is greater than the selected decision threshold, tr, then the device is identified as authorized and the type is as predicted by the classifier. Otherwise, the device is identified as unknown. We set the decision thresholds $tr = 0.6 - 0.85$ through our experiments.

Implementation and Parameters Tuning. In our experiments, we use Scapy [212], to decode network traces. These traces are then applied to the features extractor, to extract the unique selected features from a sequence of decoded packets. The set of selected features with their belonging labels feeds the classifier for training. Thus, the trained classifier can distinguish between different white-listed IoT devices.

3.6.2 CLASSIFICATION MODEL SELECTION

To choose the best classification technique for our problem, we examined a number of ML classification techniques and compared their validated accuracy. We use scikit-learn Python library [242] to implement and train the tested models. We also use 10 iterations and 10-fold cross-validation to generalize the validation results.

Figure 3.2 presents a comparison between all tested algorithms' accuracy results: LR, ABOOST, LDA, KNN, CART, NB, SVM, RFC50, RFC100, and GBOOST.

We use the default scikit-learn library parameters [242] in setting most of the tested algorithms. Table 3.1 presents the parameter settings for each of the tested algorithms.

Based on the findings, we choose Random Forest as the base classifier. Random Forest is an ensemble learning technique built on decision tree induction [244]. It has a number of advantages as per the information in the cybersecurity ML methods survey [38]. The advantages include:

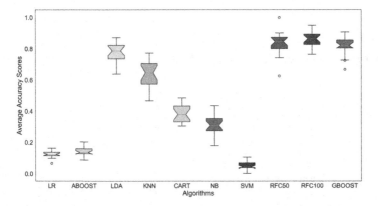

Figure 3.2: Accuracy Results For Different Classification Algorithms. RFC with 100 n_estimators achieves the best identification score with an average accuracy of 0.93 % followed by RFC with 50 n_estimators, then GBOOST with an average accuracy of 0.86%. SVM achieves the least identification score with an average accuracy of 0.1%.

- The classifier does not need feature selection.
- The algorithm is resistant to overfitting.
- Increasing the number of trees decreases the variance without adding bias.

After selecting the base classifier, we examined the two multi-class classifier types, which are the One Versus-Rest (One-Vs-Rest) classifier and the One Versus One (One-Vs-One) classifier. On the one hand, One-Vs-Rest also called the One-Vs-All classifier is a multi-class classifier that trains one classifier per class. It creates N classifiers for an N class problem. In our current solution, we have 27 different classes. Accordingly, a total of 27 classifiers will be trained. For $class_i$ it will assume (i)-labels as positive and the rest as negative, where N reflects the total number of classes and $class_i$ represent the ith classifier. This method is efficient and inoperable, as it ensures that each class is served by only one classifier. On the other hand, the One-Vs-One classifier trains a separate classifier for each different pair of classes. It creates $(N*(N-1)/2)$ classifiers [242]. Accordingly, a large number of classifiers will be created. Thus, this cause long training time, especially in solutions with a large number of classes, such as the presented scenario. The average accuracy of both techniques is very close. However, training the One-Vs-Rest classifier is faster and more practical. Accordingly, we choose to use it as the solution classifier with Random Forest as the base classifier.

The One-Vs-Rest multi-class classifier is implemented. The base classifier is Random Forest with 100 estimators (decision trees). To train the classifier, the entire dataset is converted into features and labels. The features and labels are randomized and divided into training and cross-validation sets. The shuffling

Table 3.1

Parameter Settings of Tested Algorithms.

Algorithm	Parameters
ABOOST	n_estimators= 50 learning_rate= 1 base_estimator=Decision Tree (max_depth= 1)
LDA	n_components= 10 learning_decay= 0.7 learning_offset= 10
KNN	n_neighbors= 5 leaf_size= 30 power_parameter(p)= 2
CART	max_depth= 1 min_samples_split= 2 min_samples_leaf= 1
NB	var_smoothing= 1e-9
SVM	degree= 3 decision_function_shape= 'ovo'
RFC50	n_estimators= 50 min_samples_split= 2 min_samples_leaf= 2
RFC100	n_estimators= 100 min_samples_split= 2 min_samples_leaf= 2
GBOOST	n_estimators= 100 min_samples_split= 2 min_samples_leaf= 2

process is to ensure that neither the training set nor the cross-validation set is biased by any particular device/device type. Finally, for each experiment, we log the training set and validation set accuracy, F-scores, and zero-one loss.

3.6.3   EXPERIMENTS AND RESULTS

Using the described model, the network is trained and then evaluated to distinguish between known vendor models, individual devices that have the same type, vendor and model, and unknown devices. Moreover, we investigated the feature importance of each of the selected features, on the accuracy, weighted-average precision, recall, and F1-score of the classifier.

3.6.3.1   Device Type Identification

The extracted features and labels for the pcap files for all 27 different device types are randomized. The feature set and belonging labels are divided into training and validation sets. We use the train_test_split module from the scikit-learn library to divide the fingerprints dataset, into 80% for training

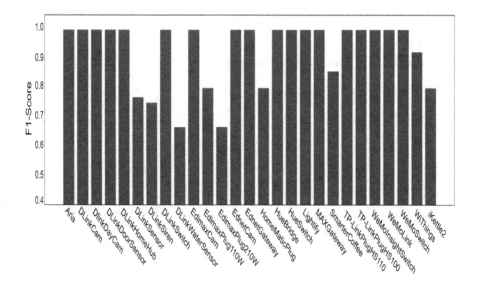

Figure 3.3: Performance of Device Types Identification.

and 20% for validation. The network is trained with the training set. The validation set is then applied to the trained network. The average recorded weighted F-score for identifying device types is 91% and a loss value of 0.09999. The F1-score for each of the device types is shown in Figure 3.3.

Device types with low identification rates such as D-LinkSensor (model: D-Link Wi-Fi Motion sensor DCH-S150), D-LinkSiren (model: D-Link Siren DCH-S220), and D-LinkWaterSensor (model: D-link Water sensor DCH-S160), are almost device types that share the same vendor. Thus, the vendor uses, mostly, similar technology and configurations, making it difficult to identify these types perfectly. Other, device types identification rates are almost perfect; for instance, Aria (model: Fitbit Aria Wi-Fi-enabled scale) is identified perfectly with an identification-weighted F1-score of 100%.

3.6.3.2 Unknown Device Identification

We conduct 27 different experiments, as per the number of IoT device types in this dataset. In each experiment, the network is trained with 26 IoT devices. Leaving one IoT device as the unauthorized device type. The threshold, tr, is set as discussed in Section 3.6.1 to optimize the F-Scores. Features are extracted from all 26 white-listed devices, then they are accompanied by their label to create the white-listed dataset. The classifier is then trained with this created dataset. Then, features are extracted from each of the pcap files for the device in question, which is the device in turn that is not in the white list. These extracted test device features are applied to the trained classifier. Then the unknown device identification rate for each of the devices

Table 3.2

Unknown Devices Experiments Parameters.

Device Name	#. Samples	IR	C. Technology
Aria	20	1	Wi-Fi
HomeMagicPlug	20	1	Proprietary Technology
Withings	19	1	Wi-Fi
MAXGateway	20	1	Ethernet
HueBridge	20	1	Zigbee & Ethernet
HueSwitch	20	1	Zigbee
EdnetGateway	20	1	Wi-Fi
EdnetCam	20	1	Wi-Fi & Ethernet
EdimaxCam	20	0.85	Wi-Fi & Ethernet
Lightify	20	1	Wi-Fi & Zigbee
WeMoInsightSwich	25	1	Wi-Fi
WeMoLink	20	1	WiFi & Zigbee
WeMoSwitch	25	1	Wi-Fi
D-LinkHomeHub	20	1	Wi-Fi & Ethernet & Z-Wave
D-LinkDoorSensor	25	1	Z-Wave
D-LinkDayCam	20	1	Wi-Fi & Ethernet
D-LinkCam	20	1	Wi-Fi
D-LinkSwitch	20	1	Wi-Fi
D-LinkWaterSensor	20	0.5	Wi-Fi
D-LinkSiren	20	0.6	Wi-Fi
D-LinkSensor	20	0.55	Wi-Fi
TP-LinkPlugHS110	20	0.35	Wi-Fi
TP-LinkPlugHS100	20	0.3	Wi-Fi
EdimaxPlug1101W	20	0.1	Wi-Fi
EdimaxPlug2101W	20	0.4	Wi-Fi
SmarterCoffee	20	0.3	Wi-Fi
iKettle2	20	0.1	Wi-Fi

Note: #. samples represent a number of tested pcap files, IR represents Unknown Identification Rate, C. Technology represents Device Connection25 Technology.

is calculated as per each experiment. Finally, the overall average accuracy in identifying unknown devices is calculated. Table 3.2 summarizes the 27 different experiments described above with a decision threshold set equal to 0.6.

3.6.3.3   Individual Device Identification

The dataset used in this experiment consists of all 31 different devices. Among these devices, there are 4 pairs of individual similar device types, which are:

- EdnetCam1 & EdnetCam2
- EdimaxCam1 & EdimaxCam2

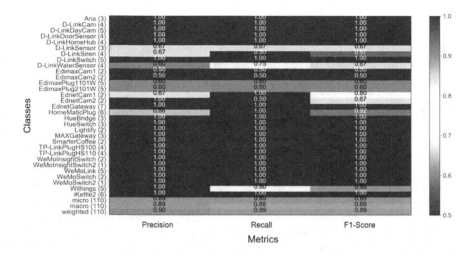

Figure 3.4: Classification Report for Individual Device Instances.

- WeMoInsightSwitch & WeMoInsightSwitch2
- WeMoSwitch & WeMoSwitch2

3.6.3.4 Individual Device Identification

The dataset used in this experiment consists of all the 31 different devices. Among these devices, there are 4 pairs of individual similar device types, which are:

- EdnetCam1 & EdnetCam2
- EdimaxCam1 & EdimaxCam2
- WeMoInsightSwitch & WeMoInsightSwitch2
- WeMoSwitch & WeMoSwitch2

Each pair shares the same device type, model, and vendor. Features are extracted from all the pcap files of all devices. The feature vector for each pcap is then accompanied by its related label. This creates the overall dataset. This dataset is split into 80% for training and 20% for validation. The trained network is then tested with the validation set to test the accuracy of the classifier. In summary, the F1-score identification rate is between 55%-100%. Figure 3.4 shows the comparison of the classification report for all tested devices.

Devices with a similar type, model, and vendor are correctly identified in more than 50% of the cases. The low identification of some of these devices is due to the availability of a limited number of pcap files for each of these individual instances. For instance, EdimaxCam2 has only five pcap files available in the dataset. These limited network traces files are split into training and validation datasets. Accordingly, the classifier is not well-trained with this device instance.

3.6.3.5 Feature Importance

Besides assessing the overall accuracy level, we inspected the features and their importance. Figure 3.5 illustrates each of the 67 features and their importance score. Importance is the total average decrease in node impurity over all trees of the classifier, which is weighted by the probability of reaching that node [44]. The most important feature is the average TTL. The second, third, and fourth important features are all related to summary statistics of Ethernet packet size, which are the maximum, mean, and variance of Ethernet packet size, respectively, followed by the TCP payload data offset feature. Moreover, the three least important features are all related to IP header size, which is median, q1, and minimum IP header size. The values of these three features are almost zero. However, we decide to keep these features, for generalization, as they might have a minor effect when using other datasets.

3.6.3.6 Comparison with Existing Approaches

The fact that we are using the same dataset as IoT sentinel directed us to compare our results with this state-of-the-art solution. We use the F-score for validation, while the authors of IoT sentinel used an accuracy metric. Accordingly, we compare here the accuracy metric results we achieved, to be able to compare the two solutions. On the one hand, the authors of IoT sentinel extracted 23 features from each packet header in pcap file to create a fingerprint for each device. They achieved an identification accuracy of 81%. On the other hand, we use 67 features from only a sequence of 20-21 packets from each pcap file. Features included both header and payload information, achieving an average identification accuracy of 90.3%. Finally, the proposed solution identified individual devices sharing the same vendor and model. The achieved identification accuracy is 0.89%. However, the IoT sentinel proposed solution did not focus on individual device identification nor did it include experiments or results to identify individual devices.

3.7 SUMMARY

This chapter presents a security enforcement model design that identifies and authorizes IoT devices using behavior flow-based fingerprints and ML. The described model demonstrates a novel IoT passive device fingerprinting technique. The unique fingerprint is created from features selected from the network packet traces, using the information in both the packet's headers and payloads. The ML network is trained with fingerprints of network white-listed

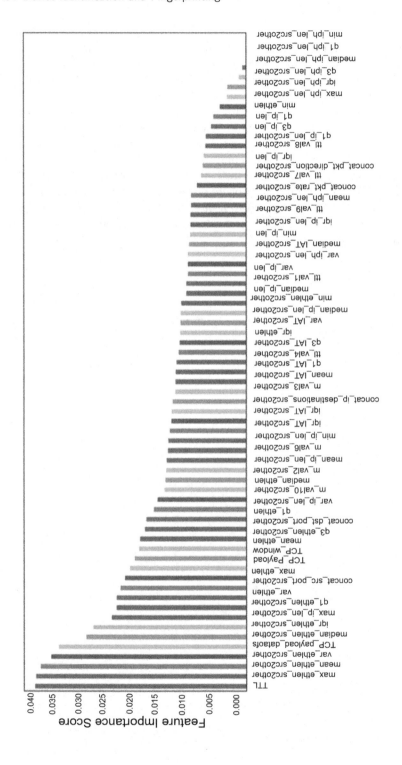

Figure 3.5: Features Importance Scores.

devices. We conduct a number of experiments, to demonstrate the proposed IoT identification approach and further showcase the effectiveness of our proposed security model design. Fingerprints that are created, feed an automated device identification that uses a supervised multi-class One-Vs-Rest machine learning network. The trained One-Vs-rest Random Forest network can distinguish between device types as well as individual device instances from the same vendor and model. The identification F1-scores achieved respectively are 91% and 89%. Moreover, the created fingerprints are stored as baselines for each device, to detect rogue or compromised devices. Privileges on an IP network are assigned to each identified device as per predefined rules. The behavioral fingerprint of random capture of each device network communication is created and then compared to a baseline. If a device is detected with behavioral changes as detailed in the next chapter, it will be quarantined for further monitoring.

4 Behavioral Novelty Detection for IoT Network Traffic

IoT-applied solutions are changing the way the world perceives technology. IoT devices are now being used in a wide range of applications to transfer or share relevant information with each other, hence reducing human interventions. With such a wide spread of IoT solutions, security becomes a major concern. Many IoT devices are vulnerable due to several reasons including insecure implementations, poor life cycle management, and inappropriate configurations, leading to an increase in the risk of these devices getting exposed and attacked. However, the current security approaches for detecting compromised IoT devices are not efficient, especially for *zero-day* attacks as noted in Chapter 2. Since no one knows what a new attack would look like, it will be useful to monitor and detect anomalies using accurate detection techniques. In Chapter 3 we discussed the effectiveness of ML algorithms identifying IoT devices within a network and we propose a security framework (BIN-IoT) that not only provides access control service but can continuously monitor the IoT devices' communications to detect any drift from their normal behavior. This chapter develops a solution that acts as the anomaly detection module in our BIN-IoT security framework named Behavioral Novelty Detection for IoT Network Traffic (BND-IoT). BND-IoT probes the possibility of detecting IoT network traffic anomalies using novelty detection techniques, thus it can detect compromised IoT devices. BND-IoT trains a neural network with novel selected behavioral features extracted from benign traffic only and then uses the novelty techniques to detect any unusual traffic patterns. We show that the presented approach is effective in detecting anomalies within the network traffic of IoT devices with a strong average F1-score of 96.7%, and a low false rejection rate of 7%. The main content of this chapter has been published in Hamad et al. [105].

4.1 OVERVIEW

Billions of IoT devices are connected to each other, and the number is still increasing day-by-day. IoT has become a considerable element in almost everybody's daily life. With the expanded deployment of IoT solutions in almost every domain, security becomes a critical issue. Most IoT devices and solutions are vulnerable to attacks, as they are neither updated frequently nor configured securely, e.g., using simple passwords or applying default configurations.

It was reported recently [36] that attackers exploited insecure IoT devices and used these devices as bots to launch volumetric DDOS and Brute-force attacks.

Controlling IoT network traffic involves monitoring the massive amount of moving data. Hence, it is easy for malicious activities to hide within the normal traffic and infiltrate the system without being detected, especially without prior knowledge of such traffic patterns [138]. One of the main goals of a security solution is to have a scalable and reliable IoT infrastructure that is capable of detecting compromised IoT devices and malicious traffic within the network. Here, ML techniques can be used to detect malicious traffic traces and protect the IoT infrastructure [110].

In recent years, ML techniques have proven their efficiency in many mission-critical applications, including anomaly and Intrusion Detection System (IDS) [238]. A number of current security solutions use learning-based approaches, in which models are trained using comprehensive big datasets [238]. These trained models can be integrated with firewalls to improve the overall network detection and prevention solutions effectively.

Our work focuses on providing a solution that can detect malicious behaviors in IoT network traffic using anomaly detection techniques. Existing solutions mostly rely on outlier-detection techniques to identify and detect intrusions. As a result, they are not ideal for real-time IoT anomaly detection (Section 4.2 discusses a number of these solutions). In this chapter, we present a solution named BND-IoT. The goal of the BND-IoT system is to detect compromised IoT devices and malicious traffic within infrastructure in real-time using novelty detection algorithms. The input of our detection model is a novel behavioral-based fingerprint, generated from normal traffic patterns only for each IoT device type. A detailed discussion of the novel fingerprinting technique and the novelty anomaly detection technique is presented in Section 4.4.

The key contributions of this chapter are as follows:

- We propose a novel behavioral fingerprinting technique that captures the behavior patterns of IP-enabled IoT devices' network traffic. This behavior-based fingerprint can be used by the novelty detection solution to catch malicious traffic, thus allowing the system to identify compromised IoT or rogue devices connected to the network.
- In the absence of any prior expert knowledge of anomalous data, we propose BND-IoT, which is a real-time IoT network traffic anomaly detection system that can detect anomalous traffic from unknown, unseen attacks, and malware traffic when the network model is trained with normal traffic only. The aim is not only to detect known attacked patterns but also zero-day attacks with high Detection Rates (DRs) and low False Positive Rates (FPRs).

The rest of the chapter is organized as follows. Section 4.2 overviews the related work on network anomaly detection. In Section 4.3, we present both the threat model and the proposed approach to detect compromised IoT devices connected to an IoT infrastructure. We then propose BND-IoT, which is an IoT device anomaly detection system in Section 4.4. This section includes details on the BND-IoT system architecture design and components. Section 4.5 reports the experimental studies and the results. Section 4.6 discusses the proposed solution and observed challenges. Finally, Section 4.7 concludes the chapter.

4.2 RELATED WORK

There is a significant amount of network traffic anomaly detection research in the literature using ML techniques. The works in Eykholt et al. [6] and Rajasegarar et al. [224] targeted anomaly detection in sensor networks and industrial control infrastructures, respectively. Several approaches have been proposed to detect and prevent intrusions or anomalies in IoT infrastructures. A number of these proposals used two-class classification (benign and malicious) [81, 110, 182, 197]. This approach opposes the objective of an anomaly-based approach that should detect any variation from the benign traffic behavior [67].

Several approaches extracted features including flow-level statistics such as counting the number of packets, and the rate of number of unique ports to train the classifiers. Lately, some DL approaches, especially Recurrent Neural Network (RNN), were proposed that were trained with both normal and anomalous traffic with a similar set of features to detect anomalies, intrusion, and malicious infected devices within a network [108, 113, 172, 179, 187, 234, 253, 288]. In contrast to these researches, our work proposes an approach that models the normal traffic behaviors of IoT devices and uses novelty detection algorithms that train the network with normal benign traffic only. In general, IoT devices have systematic behaviors, which usually connect to particular servers or do specialized tasks [30]. Accordingly, novelty detection techniques can be useful; thus, behavioral deviations can be identified, disregarding the infection types. Moreover, the classifier can be trained faster and can detect unseen (zero-day attacks) anomalies in live network traffic.

In the literature, many one-class classification techniques have been proposed to identify network traffic intrusions or anomalies. A number of these techniques utilized the Support Vector Machines (SVM) methodology [138]. Bezerra et al. [30] tested four one-class classifiers to detect anomalies in IoT networks. The best achieved F1-score was 94%. Nguyen et al. [192] proposed a one-class classification approach based on RNN and federated self-learning technique. The classifier was trained with a sequence of symbols that represent normal benign traffic only that were collected at several security gateways. Each gateway was dedicated to monitoring a client IoT network. On the one hand, the classifier in Nguyen et al. [192] was tested with anomalies generated from only Mirai malware [126]. On the other hand, this work used a dataset with generic types of attacks that can target any IoT device.

4.3 THREAT MODEL

We consider an adversary model, where malicious attackers are trying to compromise vulnerable IoT devices in the infrastructure. Attackers that compromised IoT devices can gain access to sensitive information on the compromised devices or use these devices as pivots or bots for further attacks. We also consider IoT devices in the infrastructure that are already compromised or infected with malware and produce unusual network behaviors.

A security system not only aims to preserve the privacy of IoT devices' data, but also protects any device connected to the network from getting exploited or compromised. The goal of this work is to detect attacks on IoT devices or identify infected IoT devices, thus be able to take appropriate defense or protection methods, such as prevention of malicious traffic to communicate with inter-devices or isolating compromised devices when the infection is detected.

In this chapter, we assume that the IoT devices are not compromised at the time of collecting the normal traffic. Accordingly, the normal network traffic of IoT devices collected for training the network is benign and free from any malicious traffic.

By considering the IoT infrastructure and security challenges discussed in Section 3.3.1, Chapter 3, we propose a self-learning anomaly detection technique, that learns baselines of IoT devices using devices sample traffic. The approach uses unsupervised learning algorithms to detect anomalies from unlabelled network traffic generated from IoT devices connected to the network. In particular, this proposed solution can be applied in both domestic and industrial sectors, to detect IP network traffic anomalies of connected IoT devices.

It should be noted that the solution is concerned with IoT devices that are already connected to the infrastructure and that continuously recheck their network communication traces against a baseline to detect any compromised device or malicious traffic. The technical details can be found in the next section.

4.4 BND-IoT SYSTEM

We propose BND-IoT, an IoT network traffic anomaly detection system to identify malicious traffic in the IoT infrastructure as well as compromised IoT devices within the network. To manage the challenge of high FPs in IoT anomaly detection due to the heterogeneous nature of IoT, we propose a trained classifier for each IoT device. Generating a behavioral baseline for each IoT device's normal network traffic, we then continuously monitor the network traffic for each of the connected IoT devices to detect any anomalous

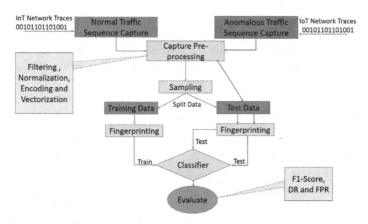

Figure 4.1: System Model for IoT Network Traffic Anomaly Detection. The diagram includes both the training and testing phases. The evaluation of the trained model can include a number of metrics, such as accuracy, anomaly DR, F1-Score, and FPR.

traffic that deviates from the generated baseline. This detected anomalous traffic could be malicious traffic trying to compromise an IoT device or could be generated from a compromised IoT device within the infrastructure.

4.4.1 BND-IoT ARCHITECTURE

The main components of the BND-IoT system architecture design include a fingerprinting solution to select and extract features and a classifier network, the details of the model architecture are as shown in Figure 4.1.

The network traces of IoT devices already connected to the infrastructure are monitored. The first component of the model is a packet capture that uses a random moving window to capture a sequence of packets from the monitored IoT devices' network traces. This picked-up traffic is then applied to a pre-processing module to prepare the data before applying it to the fingerprinting solution. Pre-processing of data ensures all missing data (Null) are handled and the categorical data are converted to numeric and normalized. Then the pre-processed data is directed to the fingerprinting solution to select and extract features. The generated fingerprints are then fed into a classification network. This classification network is trained only with benign traffic. The classifier tries to match this traffic with the stored baseline for the IoT device traffic behaviors. The decision values of the classifier can be equal to 1 for normal benign traffic, or either 0 or a negative value for unusual traffic traces. Since the classification is performed on a window of a sequence of packets, the classifier triggers the detection of an anomaly only if the selected sequence contains a considerable number of anomalous packets, thus reducing the FPs. Unusual traffic traces or anomalous traffic indicates that the device may be compromised. Hence, this device should be isolated from the network and monitored for further analysis. This further monitoring and analysis can be performed by Security Event and Information Management (SEIM) solutions.

4.4.2 FEATURE EXTRACTION

In this section, we present a feature selection and extraction scheme to generate a baseline fingerprint for each IoT device. The generated behavior-based fingerprint is used to train and test a classifier. Then the performance of the novelty detection algorithms with the new set of features is considered.

Following the proposed feature selection technique in Hamad et al. [107] as described in Chapter 3, we divide the data into windows of sequences of 30 packets for each IoT device. In Hamad et al. [107] a fingerprint for each device was created by extracting flow-based features that include header and payload information from a number of network layers and protocols, such as Ethernet, IP, and Transmission Control Protocol (TCP), and User Datagram Protocol (UDP) information. Such information was captured from a sequence of IoT network packets. A 67-dimensional feature vector was used to represent unique network traffic features for each device type. The selected features include summary statistics such as minimum, maximum, mean, and Fast Fourier Transforms (FFT) for a number of the selected features.

In this chapter, we capture bidirectional device-device communication packets, while using the Media Access Control (MAC) address to identify the source and destination of packets. We generate a fingerprint for each device type using both behavioral and flow-based features as depicted in Figure 4.2 [105]. On the one hand, to capture the behavior of each IoT device, we include information such as ports used and the usage frequency of these ports [182]. On the other hand, to capture information flow-related data, we include information such as IP destinations and a number of traffic statistical information, such as Inter-Arrival-Time (IAT) [182]. To measure variability in data, we use statistical quartiles. Quartiles (q) are values that divide an ordered data into quarters [214]. Specifically, q1 is the calculation of the middle of the lower half of the data, q2 is the median and q3 is the calculation of the middle of the upper half of the data [214]. We included some of the features from the behavioral features set in Chapter 3 as follows:

- Summary statistics of Ethernet packet size, IP packet, header sizes, and payload size.
- Summary statistics of TCP payload data offset. Data offsets communicate the data start point to the upper network layers. Hamad et al. [107] presented the significant effect of this feature for identifying IoT devices.

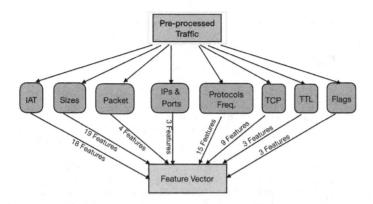

Figure 4.2: BND-IoT Feature Selection and Extraction. The pre-processed traffic is passed to eight modules to select and extract behavioral and flow-based features. IAT features include statistical data and FFT for IAT. Sizes include Ethernet, IP and payload length features. Packet features consist of packet-related information including, packet order and frequency. IP destinations, source and destination ports are selected and extracted in the IPs and Ports module. Five protocols' frequency information is extracted in the protocols frequency module. TCP-related information is extracted by the TCP module. The TTL module extracts and calculates TTL statistical information. The flags module extracts statistics of packet flags.

- TCP window size feature. TCP window size represents the number of packets that are sent and acknowledged in a single acknowledgment by the receiver. This feature considers the device memory and processing speed [29].
- Summary statistics of Time-To-Live (TTL) for TCP and UDP packets. TTL is the allowed hops count for each packet. In other words, this feature presents the lifespan of packets in a network.
- IAT (Packet Inter-Arrival Time) summary statistics and 10 highest magnitudes of FFT. IAT is the measure of the delay between the following packets.
- A count of the number of IP destinations, source and destination ports within the selected window.
- Packet rate. It is used to count the number of transmitted packets in a one-minute period.

In this chapter, we add a number of other selected features to uniquely present continuous network traffic for each IoT device. We perform a number of experiments to choose the number of packets in the captured sequence. Accordingly, we choose to capture a sequence of 30 packets from the network traffic of each IoT device. The proposed feature vector contains 74 features. These selected features are described in Table 4.1.

Table 4.1

Fingerprints Features.

Number	Feature
0-17	Statistics (min, max, q1, median, mean, q3, var, and iqr) of Packet Inter-Arrival Times (IAT) and Fourier Transforms (FT) of IAT. IAT summary statistics and IAT 10 highest magnitudes of FFT. IAT is the measure of the delay between the following packets.
18-32	Statistics (min, max, q1, median, mean, q3, var, and iqr) of Ethernet Frame and Packet Header Size. Statistics for the Ethernet frame and packet header size of transmitted packets in both directions. Variations in size can detect volumetric attacks.
33-48	Statistics (mean, min, and max) of TCP Frequency, UDP Frequency, DHCP Frequency, and DNS Frequency.
49	Packet Rate. It is used to count the number of transmitted packets in a one-minute period, which is useful in detecting flooding attacks.
50-55	Statistics (mean, min, and max) of Packet Order and Payload Length. A count of the number of transmitted packets in each direction in a one-minute period. Such info can help in identifying amplification attacks.
56-58	Periodic IP Destinations, Source, and Destination Ports. A count of the number of IP destinations, source, and destination ports within the selected window. These features can indicate an anomalous connection, especially in Distributed Denial of Service (DDOS) attacks.
59-61	Statistics (mean, min, and max) of Packet Time-To-Live (TTL). TTL for TCP and UDP packets. TTL is the allowed hops count for each packet. In other words, this feature presents the lifespan of packets in a network.
62-67	Statistics (mean, min, and max) of TCP Window Size and TCP Payload Size. TCP window size represents the number of packets that are sent and acknowledged in a single acknowledgment by the receiver. This feature considers the device's memory and processing speed.
68-70	Statistics (mean, min, and max) of TCP Data offsets (Dataofs). Summary statistics of TCP payload data offset. Data offsets communicate the data start point to the upper network layers. Hamad et al. [107] presented the significant effect of this feature for identifying IoT devices.
71-73	Statistics (mean, min, and max) of Packet Flags. Summary statistics of TCP flags. The frequency and statistics of occurrence of each flag can indicate the presence of a number of attacks, such as probing attacks.

4.4.3 CLASSIFICATION NETWORK

We formalize the problem of anomaly detection classification as a *one-class classification* problem [26], as we are training the network with benign traffic and expecting the network to classify network traffic as one for benign traffic and zero or negative for anomalous traffic depending on the type of classifier.

There are different approaches including Deep Learning techniques that can be used as a base for performing one-class anomaly detection classification, specifically novelty detection for IoT network traffic. In this chapter, we are interested in examining four different approaches for novelty detection, namely, Isolation Forest (IF) , Local Outlier Factor (LOF) , Elliptic Envelope (EE) , and RNN architectures.

We set novelty to true for IF, LOF, and EE by setting the contamination to zero for IF and EE classifiers and setting the novelty parameter to true for the LOF classifier. In the RNN classifier, we use a fully connected layer followed by a sigmoid regression layer. The dimension of the fully-connected layer is set equal to the feature dimension. The sigmoid layer has two outputs to represent normal traffic and abnormal traffic. All classifiers are trained with normal traffic to ensure classifiers learn benign traffic only. We expect that Long Short-Term Memory (LSTM) can also provide better results than RNNs. This is due to the fact that LSTMs control the vanishing gradient issue faced by RNN [101]. However, defining a time step as well as a base to identify unusual behavior is an issue for this technique.

IF classifier presents the best results for most of the devices in comparison to the other three approaches. An in-depth analysis of these experiments can be found in Section 4.5. IF produced almost ideal anomaly Detection Rates (DRs) and low False Positive Rates (FPRs). IF builds a collection of binary isolation trees that can identify outliers or anomalous data instead of normal data. It has the ability to scale up to identify anomalies in big data as they have low memory requirements [140]. Anomaly instances usually have short average path lengths in comparison to the normal instances [154]. IF separates anomaly instances which are usually closer to the root of the tree, while the normal traffic instances are more likely to be at the deeper ends [154]. Anomaly score S(x,n) for each instance reflects the possibility of being an anomaly and is calculated using the average path length from the IF trees as described in Equation 4.1 [153]:

$$S = \frac{2.E(h(x))}{c(n)}; \quad E(h(x)) = \frac{1}{L} \cdot \sum_{i=1}^{L} h_i(x) \tag{4.1}$$

Where x is the test sample, L indicates the number of trees in the forest, $E(h(x))$ is the calculation of the average $h(x)$ for the ensemble of trees, n is the number of sub-sampling sizes, $c(n)$ is the calculation of the average $h(x)$ for a given n and h_i represents the length of the ith tree.

4.5 IMPLEMENTATIONS AND RESULTS

This section presents the analysis and evaluation of the proposed BND-IoT system.

4.5.1 DATASET

In our experimental studies, we choose to use the UNSW IoT analytics dataset [108]. To the best of our knowledge, it is one of the latest and the most comprehensive IoT traffic datasets publicly available at the time of this writing.

The dataset includes collected and synthesized network traffic traces for 30 different IoT devices connected to the UNSW lab infrastructure [108]. Normal traffic flow is collected in 26 different packet capture files (pcap files). While running the attacks experiments, Hamza et al. [108] collected 17 different pcap files that capture both attack and normal traffic traces. We use the subset that is released for community use. This community use subset contains normal and attack traffic of nine different IoT devices. We refer the readers to Hamza et al. [108] and UNSW-Sydney [275] for a detailed description of the dataset.

To use the UNSW IoT analytics dataset, we first filter the traffic for each device separately using both device's MAC address and IP address. These filtered pcap files for each device are then separated into normal traffic and attack traffic traces. The pcap files that have the network traffic traces are enormous in size, accordingly pre-processing is needed (e.g., read, split files per session, normalize) to enhance the massive amount of network traffic. We select features from the normal traffic traces to reduce the size after vectorizing the dataset to 74 features by using behavioral fingerprints for each device. Those features represent unique network traffic behavior fingerprints for each device. The extracted features are then processed and serialized [218] to be saved into files to enhance the management of the data. The processed data is divided into training and validation sets to be used for learning and prediction of normal behaviors, respectively. The same procedure is applied to attack traces and the extracted behavior fingerprints are saved in separate files to be used for testing and evaluation.

4.5.2   EXPERIMENTAL SETUP

We use the network traffic from nine different IoT devices. We train the neural networks with normal traffic filtered to detect the traffic of each device separately. We develop our attack detection solution using Python programming language libraries (e.g., Pandas, Numpy, sklearn, keras). We divide the benign traffic using the train-test-split module from scikit-learn library [243], into 80% for training and 20% for validation. The attacked traffic dataset is used for testing the accuracy of the proposed solution, while the validation normal traffic is used to identify the False Positive Rate (FPR).

Classification Model Selection. To utilize the IF, LOF, and EE classes, we import them from scikit-learn [243]. Regarding RNN class, we import it from keras [133]. We use the default scikit-learn and keras parameters [133, 243] when setting most of the parameters of the selected algorithms. Parameters for each classifier are presented in Table 4.2. We run a number of additional experiments for other novelty detection classifiers, such as one-class SVM. However, their results are not good enough to be included in this work.

Evaluation Metrics. This work is presented as a one-class classification problem based on binary classification to classify data as normal or attacked. Detection rate (DR) is a representation of correctly identified anomalies or can

Table 4.2

Parameter Settings of Tested Algorithms.

Algorithm	Parameters
IF	n_estimators = 250 behaviour = new max_samples = 5000 random_state = rng contamination = 0
LOF	n_neighbors = 100 algorithm = auto novelty = True
RNN	Network activation = relu except for the last layer with activation = sigmoid. First layer is with input size and last layer output = 2. optimizer = rmsprop loss = binary_crossentropy metrics = accuracy
EE	store_precision = True contamination = 0 random_state = 0

also be called True Positives (TP), True Positive Rate (TPR), sensitivity, or recall. To detect the percentage of the incorrect classification of normal traffic as malicious, we utilize FPR. Our objective is to increase the accuracy while minimizing the False Positives (FPs); otherwise, the system will not be practical for real-time monitoring and alerting. The DR for the anomaly detection and its relationship with TPs and FNs [267] is represented in Equation 4.2.

$$DR = \frac{TP}{TP + FN} \tag{4.2}$$

The confusion matrix and other metrics (classification report and Receiver Operating Characteristics (ROC)) are used to show the accuracy (TP and True Negative (TN)) and correctness of the proposed model. The confusion matrix presents normal network traffic traces that are either correctly (TN) or incorrectly (FP) classified as well as anomalous traffic traces that are correctly (TP) or incorrectly (FN) classified. The confusion matrix for one of the devices is shown in Figure 4.3.

The precision metric shows how many of the detected malicious traces are correct, and the recall metric illustrates how many of the malicious attacks the model detects. Furthermore, the F1-score metric can gather both precision and recall metrics to show the average. The relationship between the correctly and incorrectly identified instances with F1-score is shown in Equation 4.3.

$$F_1 - Score = \frac{2.precision.recall}{precision + recall} = \frac{2.TP}{2.TP + FP + FN} \tag{4.3}$$

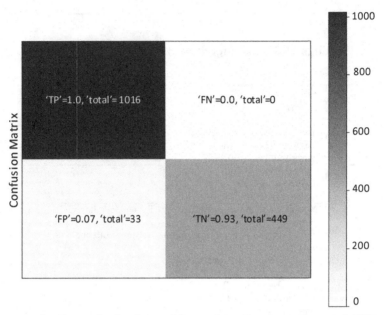

Figure 4.3: An Example Confusion Matrix for a Samsung Camera IoT Device. The confusion matrix is calculated to present the accuracy of the classifier. In each block, the total value represents the total number of samples that are classified.

4.5.3 PRELIMINARY RESULTS

We conduct the performance evaluation of our anomaly detection solution first considering IoT devices individually and then all together.

The performance results (F1-score) for different novelty detection techniques are shown in Figure 4.4. The results show that IF classifier performs better than other classifiers.

We are able to detect 99.9% of all anomalous or malicious traffic using IF, in other words, achieving a TPR of 99.9%. Figure 4.5 demonstrates the ROC curve of FPR and TPR for all IoT devices. The figure shows 95% or higher TPR while maintaining a low FPR, which is one of the goals of our work. Figure 4.6 demonstrates the IF classification report for all IoT devices. The classification report shows the precision, recall, and F1-score for each IoT device. The overall F1-score achieved is 96.7%.

We use the validation data for calculating FPR. Any trigger for the anomaly in this data will indicate an FP as the validating data contained only benign traffic. Figure 4.7 demonstrates the DR vs. FRR for all devices. The malicious traffic DR for most of the devices is almost perfect except for the Wemo motion sensor which has anomaly detection of 99%. Regarding the FP, we

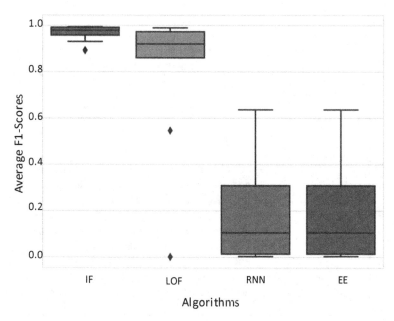

Figure 4.4: Comparison between the Four Novelty Techniques when Tested with the Nine IoT Devices. On the one hand, IF is the best achiever with an average F1-score of 96.7%. On the other hand, both RNN and EE produce the lowest F1-scores with an average result of 40%.

achieved 2-7% FPR. Knowing that in real IoT network settings that include a massive number of connected devices, this FPR may not be very efficient. The higher the FPs, the more false alerts are generated for the infrastructure administrators to investigate.

4.6   DISCUSSIONS

This section discusses the challenges that we observed during our experimental studies with the proposed system.

One of the early challenges that we face in this work is finding a suitable dataset that perfectly represents actual IoT network traffic with abundant training and testing data. Most of the available datasets use either synthetic IoT data or do not include IoT network traffic. Accordingly, we decided to use the IoT dataset from [275] to prove our concept.

Observations. In our experiments the FPR for most of the IoT devices is low, except for the TP-Link plug and the iHome devices. The difference in the FPR for these two devices encourages us to analyze the dataset and investigate the reasons for the higher FPRs. The Huebulb and the Netatmo camera are

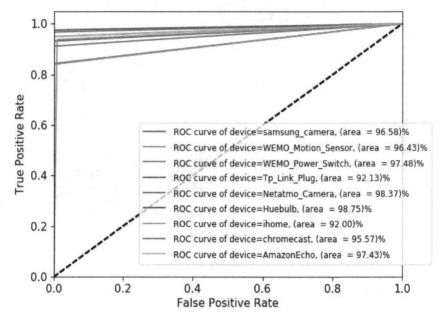

Figure 4.5: Receiver Operating Characteristics (ROC) Curve for the IoT Devices using IF Classifier. The ROC curve shows the TPR and FPR for each of the IoT devices. The area under each of the curves is also calculated and illustrated.

observed to have the lowest FPR. By analyzing the training data used for these devices, we find out that these two devices have more training data instances in comparison with the training data instances used for the devices that produce high FPRs. The benign instances that we use for training and testing the FPs on the four devices are as below:

- 634 and 147 instances for training and testing the FPs of the Tp-link plug and ihome classifiers, respectively.
- 3,966 and 9,171 instances for training and testing the FPs of the Netatmo camera and Huebulb device classifiers, respectively.

We expect that including more training data for these devices can reduce their FPRs.

Scalability. This work dedicates a trained classifier for each type of IoT device, to ensure that the proposed solution is scalable and independent of the number of IoT device types connected to the infrastructure. The model focuses mainly on the network behavior of each device type. This can help reduce the FPs and increase the anomaly DR. Adding any new type of IoT devices to the current infrastructure, will include adding a new trained classifier dedicated to the new device type. This will not affect the performance of the other

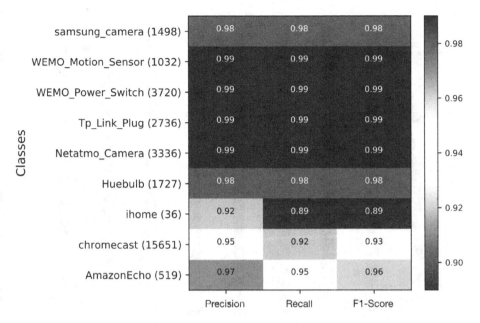

Figure 4.6: Classification Report for the IoT Devices Using IF Classifier. The highest F1-score is 99% for the Wemo motion sensor, Wemo power switch, Tp-link plug, and Netetmo camera classifiers. While the lowest F1-score is 89% for the ihome device classifier.

classifiers within the infrastructure.

Generalization. The proposed model is evaluated on different types of attacks that can be caused by different types of malware. The results show a very high detection rate of malicious traffic since the proposed system isolates any deviation from the generated IoT network traffic fingerprints. Moreover, the generated fingerprints are based on behavior information independent of the communication protocols and the dataset used, which makes the fingerprinting technique useful for different datasets and environments.

Comparison with other techniques. Our work considers more details of the IoT fingerprinting process and steps used as well as the pre-possessing of data in comparison to other works. Moreover, our work relies only on normal traffic traces to train the classifier, making it possible to identify any unseen malicious traffic. There are other works in the literature that target a one-class approach to identifying anomalies. Ideally, a logical comparison should happen between models that use the same dataset and/or the same evaluation metrics. The state-of-the-art approaches use unpublished proprietary IoT datasets. However, we discuss the differences between the state-of-the-art approaches and our approach, by using a publicly available IoT dataset. On the one hand,

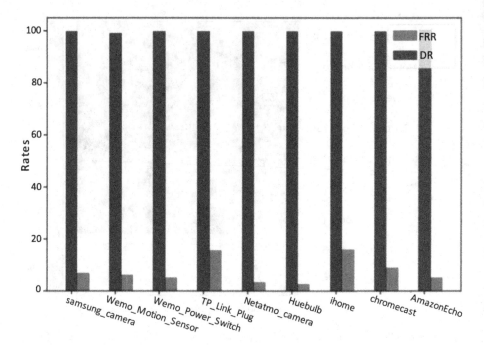

Figure 4.7: Anomalous Traffic DRs vs. False Rejection Rates (FRR) for All Devices. The dark-colored bars represent the malicious DR, showing almost perfect detection performance. The light-colored bars represent the FPR.

Nguyen et al. [192] extracted features from each network packet and mapped them into symbols from a sequence of 250 packets to create a fingerprint for each device, and used Gated Recurrent Units (GRUs) [133] for identifying anomalies. The dataset included malicious traffic generated from one type of malware (Mirai) [126]. They achieved a 94.1- 95.6% DR and 0-1% FPR. On the other hand, we exploit 74 features from only a sequence of 30 packets to create a unique behavioral fingerprint. The fingerprint is then applied to a novelty detector, achieving an average 99.9% TPR and an average 7% FPR.

4.7 SUMMARY

This chapter presents a solution for a conceptual framework to encourage discussions among the security community to direct more research efforts for enhancing the security of emerging IoT infrastructures. We propose the BND-IoT system to detect malicious traffic within an IoT infrastructure. Our proposed BND-IoT system is based on a novel behavioral feature selection scheme along with novelty detection techniques. More specifically, we exploit ML and DL algorithms as a base for detecting malicious or unusual IoT device communications using novelty detection techniques. The main objective of using novelty detection techniques is to detect unseen malicious traffic, even without training the network for similar anomalous traffic.

One of the main challenges is to keep the number of false positives low, ideally at 0%, as otherwise, network administrators will not trust the system's results. Novelty detection techniques are supposed to achieve this goal [111, 167, 279]. However, our analysis indicates that it may be more difficult to achieve in the detection of IoT network traffic anomalies due to several unique challenges. The results demonstrate that our BND-IoT system can achieve 99.9% anomalous traffic detection and an average of 7% false rejection rate.

5 Detection of Malware Targeting Embedded Devices

Detecting malware in embedded systems is becoming a critical task in various IoT infrastructures in which embedded devices perform crucial tasks (e.g., monitoring expensive engines). The complexity of this duty increases with the recent growth of malware variants targeting different embedded CPU architectures. Traditionally, ML techniques require feature engineering before training, which requires domain knowledge that might not be applicable to unseen malware variants. Recent DL approaches have performed well on malware analysis and detection, eliminating the need for the feature engineering phase as noted in Chapter 2. Chapter 4 is concerned with detecting malicious traffic on the network layer within the IoT infrastructure. In this chapter, we propose DeepWare , a real-time cross-architecture malware detection solution tailored for embedded systems. It detects malware on the application layer by analyzing the binary file's executable OpCodes sequence representations. In particular, we use BERT embedding, the state-of-the-art NLP approach to extract contextual information within an executable file's OpCode sequence. The sentence embedding output generated from BERT is fed into a hybrid multi-head CNN-BiLSTM-LocAtt model that realizes the resources of the CNN and BiLSTM with added benefits from the local attention mechanism. The proposed model extracts the semantic and contextual features and captures long-term dependencies between OpCode sequences, which helps improve the detection performance.

We train and evaluate the proposed DeepWare[1] using datasets created for three different CPU architectures. The performance evaluation results show that the proposed multi-head CNN-BiLSTM-LocAtt model produces more accurate classification results with higher recall and F1-scores than both traditional machine learning classifiers and individual BiLSTM models. The main content of this chapter is from Hamad et al. [104].

5.1 OVERVIEW

The advancement and widespread use of modern embedded systems such as IoT infrastructures have prompted cybercriminals to devise dangerous and refined attacks against embedded devices [19, 106]. Adversaries are now increasingly focusing on the misuse of low-level vulnerabilities to infect the target systems. Meanwhile, malware detection and prevention remain active research

---

[1]https://github.com/H-S-A/Embedded_ELF_Malware (anonymous).

areas [277, 301]. However, due to the resource-constrained nature of IoT device hardware and the customized operating systems, the available malware detection methodologies for embedded systems are not tailored for real-world deployment. There is thus an urgent need for specific malware detection of embedded systems.

Traditional signature-based malware detection methods depend on the accumulation of signature libraries and malware analysts' expertise. It is challenging to accommodate embedded malware growth. On the other hand, ML techniques, and data analysis methods have recently been extensively utilized and deployed to provide automatic and efficient pipelines for solving malware detection problems [229].

Dynamic and static approaches are the two main approaches for malware analysis and automated detection techniques. Both methods can utilize machine learning techniques to detect malware. The dynamic approaches [124, 213] monitor the execution of binary executables to see abnormal behaviors. However, run-time monitoring has several common potential weaknesses, such as requiring isolation, substantial resources, and the high overhead in execution run-time [229]. Moreover, executing the malicious binary files could infect the physical environments [196]. Regarding IoT dynamic malware detection, the most inherent disadvantages are due to diverse microcontroller architectures (e.g., MISP, ARM, PowerPC, x86) [114] and resource constraints of IoT devices. Hence, it is not easy to correctly set up the required environment for running and monitoring the IoT executables. On the other hand, the static method analyzes and observes the binaries' structure without executing them. Thus, it is not necessary to build a specific run-time environment for each architecture, and the detection time is usually faster than dynamic malware detection methods. However, static analysis performance can be affected by code obfuscation [213].

Most of the recent IoT-developed malware are descendant or evolved variants of Bashlite [283] and Mirai malware [32, 95]. Consequently, IoT malware families have similarly implemented source codes and operational functions. Based on such similarities, there are several static features of the IoT malware, such as Executable and Linkable Format (ELF) structure, strings, function call graph, and OpCode that could be used to detect malicious code. Generally, to classify a binary file, we need to understand the programs' functionalities and behavior patterns. OpCodes are static discriminating features that can represent behavioral patterns of a binary file [301]. In the assembly language, an OpCode is a single instruction command to be executed by the processor. OpCode sequence is a high-level static malware detection feature extracted from the binary file by a disassembler [220, 228] that usually comprises the program's execution logic [53].

Scalable, automated malware detection systems must efficiently detect malware without human expert analysis, and machine learning techniques could best fit this need. DL is a sub-field of machine learning that uses deep neural networks to provide powerful statistical models. Popular DL models, such as CNN and Long Short-Term Memory (LSTM), have become prevalent due to their ability to learn features through deep multi-layer architecture with almost no need for data feature engineering [103]. CNN can learn essential signals from the raw data directly, retain an internal representation of the data, and does not require domain expertise to select the best features. LSTM, on the other hand, tends to remember long sentences of input data; thus it is often used as the encoder to create a contextual summary for the right prediction. However, LSTM does have flaws in remembering very long and complex sequences. Moreover, general deep neural networks cannot highlight some of the input OpCodes' importance more than others during predictions.

The attention mechanism is one of the most recent valuable DL inventions. Recent innovations in NLP, such as Transformers and Google's BERT architectures,[2] are based on attention. The attention mechanism can generate weight values for different words in a sentence, giving higher weights to the core words. Since the OpCode sequence can be interpreted as a sentence, with each OpCode acting as a word in the sentence, the attention mechanism can be used with OpCode sequences. Therefore, the attention mechanism can be combined with the DL models to extract critical information describing the context and the semantics of the OpCode sequence. The use of attention-based embedding and a custom attention-based BiLSTM allows the model to benefit from understanding the whole data context and extracting useful signals to improve the prediction task.

The transformer model [117] comprises an encoder-decoder architecture model based on attention mechanisms to transmit an entire picture of the whole sequence to the decoder at once. BERT uses bidirectional transformer-based encoders and employs masked language modeling to generate word embedding. The generated word embedding includes the word's contextualized meaning, thus making the learning process more effective. Embedding is a low-dimensional description of a point in a higher-dimensional vector space. Similarly, BERT word embedding is a compact vector representation that translates a word to a lower-dimensional vector space to display the semantic significance and contextualized meaning of this word in a numeric form, thus allowing mathematical operations on text input.

In this chapter, we propose DeepWare , a malware detection solution that detects malware targeting embedded systems by analyzing the executable Op-Code sequence representations. We learn a malware representation using an embedder, an encoder, and a predictor. The embedder is composed of the state-of-the-art BERT embedding model [71] to capture relations within a single OpCode sequence and among the OpCode sequences in a trail. The encoder and the predictor construct a DL network named multi-head CNN-BiLSTM-LocAtt. The encoder incorporates three neural layers: a multi-head

---

[2]https://github.com/google-research/bert

CNN layer that learns features and internal representations of the embedded data sequence directly without feature engineering. This is followed by a BiL-STM layer, designed to preserve the ordinal position of an OpCode sequence in the trace, and a Local-Attention mechanism (LocAtt) [21] to compare interrelations among the different places of an OpCode sequence. Finally, the predictor consists of a dense layer and an activation layer that classify the input as either malware or benign executable.

In this work, we consider three of the extensively used embedded processor architectures [65, 246]: (1) *ARM processor*: largest current embedded IoT devices are based on top of an ARM processor. Hence, we have mainly chosen an ARM-based platform for all the evaluations; (2) *PowerPC processor*: it is used in many embedded solutions that can achieve one task in a single Central Processing Unit (CPU) cycle. They are very popular among manufacturers of System-on-a-Chip (SoC) and network devices, such as routers and switches.

Recently, IoT malware has been discovered to target these network devices [136]; (3) *MIPS processor*: it has a modular architecture, and its Instruction Set Architecture (ISA) has recently become open-source, thus increasingly being used in IoT devices [60].

Since a large number of embedded platforms use Linux as their Operating System (OS) [130], we confine our work to detecting Linux-based malware. To validate and evaluate our claims, we perform exhaustive experimentation focusing on the Linux environment with the three different types of processors on our collected datasets, which include 549 ARM, 877 PowerPC, and 749 MIPS-embedded malware and benign samples. To the best of our knowledge, this is the first malware solution that uses DL networks and BERT architecture on OpCode sequences for detecting malware that targets embedded systems. The main contributions of this chapter are as follows:

- We propose DeepWare, a real-time cross-architecture malware detection solution for embedded devices. DeepWare can run on either network devices or edge devices to detect malware. We validate our method by collecting OpCodes datasets and running all the experiments on three of the most commonly used embedded microcontroller architectures: ARM, PowerPC, and MIPS.
- We propose to combine BERT sentence embedding and CNN together with BiLSTM DL techniques for embedded malware detection. We use the OpCode BERT embedding sequence to extract the fine-grained inherent connection and the semantic behaviors between embedded programs. These extracted features are then fed into a deep learning network based on CNN-BiLSTM , which considers the hidden malicious program behaviors to detect malware.
- We construct a malware identification network of multi-headed CNN and BiLSTM networks for mining correlation and context information between different OpCodes, and a LocAtt mechanism to readjust the importance (weights) of the correlated features from Op-

Codes and their sequences to improve the deep learning ability to detect malware. We name the DL model multi-head CNN-BiLSTM-LocAtt. The LocAtt is seamlessly compounded with CNN-BiLSTM since the LocAtt weights for feature re-adjusting are calculated from both features and the hidden states of the CNN-BiLSTM. To the best of our knowledge, this is the first time a LocAtt mechanism is used for embedded devices' OpCode sequence malware detection.

- Finally, we publish an extracted and normalized OpCode dataset of IoT malware and benign applications for the three most common microcontroller architectures as a subsequent contribution.[3] We believe this dataset can contribute significantly to the development of the emerging embedded systems malware detection research area.

The rest of the chapter is organized as follows. Section 5.2 provides an overview of recent IoT malware research. We then present our DeepWare solution, an automated cross-architecture OpCode sequence-based malware detection for embedded devices in Section 5.3. This section includes the details on the proposed system architecture, the intuition behind choosing each component of the model, dataset creation, feature extraction, and modeling details. Section 5.4 reports the experimental studies and discusses the results of the proposed model. The results compare the proposed model and the state-of-the-art IoT malware detection techniques, and baseline machine and deep learning classifiers. Finally, Section 5.5 concludes this chapter.

5.2   RELATED WORK

Numerous research works employing static, dynamic, hybrid, or memory analysis approaches have been conducted to analyze and detect malware or classify malware families [130, 170, 211, 213]. Many studies review, and provide comparisons between static and dynamic strategies, stating the advantages and disadvantages of both techniques [24, 189]. Several studies choose to use the static analysis method to investigate the overall malware composition without executing it [229, 266, 301].

Detecting malware in embedded systems is an emerging research topic, especially after Mirai's attacks [126] or its malware families. ML and DL methods have been used in several static and dynamic analysis and detection methodologies. The conclusion of the aforementioned surveys and literature find that OpCode-based features are beneficial for malware detection, and there is currently a limited number of studies in the literature that target embedded or IoT devices malware detection.

Recently, DL models have been adopted to detect malware based on Op-Codes for ARM-based IoT applications [19, 65, 103, 211]. HaddadPajouh et

---

[3]https://github.com/H-S-A/Embedded_ELF_Malware

al. [103] used the OpCodes feature and RNN to detect malware in IoT with a Detection Rate (DR) up to 98.18%. Su et al. [263], converted the application binaries to gray-scale images and then used a CNN layer to detect and classify malware. They classified malware with an accuracy of 94.0%. Sharmeen et al. [252] examined static, dynamic, and hybrid analysis methods for industrial IoT malware. They showed that the permission list, API call list, and system call list are significant features for distinguishing mobile malware. Darabian et al. [65] counted the number of occurrences of ARM-based OpCodes to detect malware. They stated that the frequency of specific OpCodes in malware samples was higher than that of benign binaries. Dovom et al. [75] applied fuzzy and fast fuzzy pattern tree methods on ARM-based OpCode sequences to detect and classify malware. They achieved almost perfect classification results.

Most of these studies used OpCode sequences and deep learning approaches, producing efficient and encouraging malware detection results. Nevertheless, there is an extent to reform the current malware detection models for different embedded systems architectures. Embedded devices are implemented with distinct CPU architectures and processors, such as MIPS, ARM, PowerPC, and SPARC. Most of the current works focus mainly on the ARM instruction set [19, 65, 103, 211]. There are differences between the instruction sets on different architectures. Therefore, there is a necessity to establish an advanced cross-architecture static malware detection covering multiple architectures with limited processing time.

Moreover, many of the current OpCode sequence-based approaches extract n-grams from a sequence, which preserves some of the sequence information to a certain extent but unfortunately loses considerable code semantic information [53].

5.3 DEEPWARE: A MALWARE DETECTION SOLUTION

We propose an automated real-time malware detection solution, named Deep-Ware, for cross-architecture embedded systems. This section presents the proposed solution. We first introduce an overview of the DeepWare solution architecture. Afterward, we offer a detailed explanation for each of the building blocks of the proposed approach. Then we give details of our created datasets and all the steps taken to extract features. Finally, we describe the model layers and the data flow within each layer.

5.3.1 OVERVIEW DEEPWARE

The overall architecture of our DeepWare malware detection solution is shown in Figure 5.1. Firstly, each sample's OpCodes sequence is extracted and applied to a fine-tuned and pre-trained BERT model to extract OpCode sequence embeddings. The extracted embedding preserves the contextual features of each OpCode in the sequence and transforms these features into a

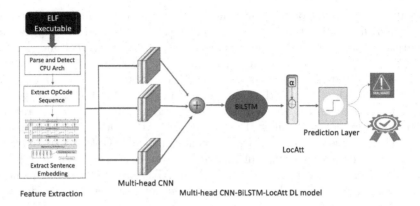

Figure 5.1: The Architecture of Our DeepWare Malware Detection Solution.

numeric form. The extracted features are then applied to a model consisting of a multi-head CNN-BiLSTM with local attention to extracting semantic information and correlations from the OpCode sequences. The model is trained and validated with OpCode sequence embeddings. After training the model, DeepWare can identify whether a binary file is malware.

5.3.2 INTUITION BEHIND THE APPROACH

5.3.2.1 BERT

Several adjoining OpCodes customarily establish a set of assembly operations with a particular context. Collecting these specific sets and analyzing them can be very beneficial in assessing samples and detecting malware. Many scholars used varying slice n-grams to capture the associated relationship between subsequent OpCode sets. However, these approaches do not usually account for a meaningful set of instructions [201]. BERT uses deep bidirectional transformers that employ a self-attention mechanism to trace relationships between all words in a sentence [278]. The sequence of OpCodes of an ELF binary file can be treated as a sentence, where each OpCode denotes a word (token). BERT uses the whole context and strives to predict the masked words' original value as per the OpCodes' context in the sequence. Accordingly, BERT pre-trained models can adequately create relationships between OpCodes in different contexts.

BERT looks at the sequence in both the left-to-right and the right-to-left paths to form bidirectional representations [201]. We can fine-tune pre-trained BERT representations for custom classification tasks. Subsequently, we use a fine-tuned BERT embedding's attention mechanism to capture an appropriate OpCodes sequence slice size and detect the selected sequences' global context.

The BERT embedding layer captures the critical OpCode information from the sample input and can efficiently learn the representation of relevant sequence patterns of the embedded systems malware. This layer maps the Opcode or OpCode sequences to an embedding vector, reflecting the semantic distance and the relationship between OpCodes. The embedding output generates different score values for different input values according to the global context [53].

5.3.2.2 Hybrid Multi-Head CNN-BiLSTM with Attention

Combining LSTM and CNN models has been used in various domains such as NLP and anomaly detection with remarkable results [41, 122]. The core intuition behind stacking CNN with BiLSTM is to respect the local information in OpCode sequences and long-distance dependencies between OpCodes [122].

The multi-head convolution is a CNN that processes each input on an entirely self-governing convolution, known as convolutional heads. The multi-head CNN is responsible for extracting significant features from the input data. We consider a multi-head CNN layer with varying kernel sizes that moves vertically down for the convolution operation considering several OpCodes from the sequence together depending on the selected filter size. Otherwise, the extraction of features would be done on the whole input sequence, leaving each state's potential vital features uncaptured. The multi-head CNN layer considers words in close-range only. Hence, it does not consider the overall context in a particular text sequence [2]. On the other hand, LSTM is a type of RNN that can remember information for an extended period. LSTMs can remember previous information using hidden states and connect it to the current task, allowing such networks to consider the long-term semantics of the sequence.

We choose to use the BiLSTM as it accesses both forward and backward information and captures the contextual information in both directions [216]. BiLSTM and its variants consider the sequence structure, but it does not usually give higher weight to more meaningful words. Since not all OpCodes have equal importance in the OpCode sequence, we use the local attention mechanism to create an aggregated representation of the important OpCodes in the horizontal sequence and extract the internal spatial relationship between OpCodes. Finally, we use a classification layer with two possible outputs: 1 in case the given sample is malware or 0 for benign samples.

5.3.3 DATASET CREATION AND FEATURE EXTRACTION
5.3.3.1 Dataset Creation

In our research study, we create three datasets, a dataset for each of the three most commonly used microcontroller architecture in embedded systems [65]. In the ARM dataset, we utilize the IoT malware and benign application dataset created by HaddadPajouh et al. [103]. The ARM dataset includes 247 IoT malware and 269 IoT benign Debian package files of 32-bit ARM-based

IoT applications. Regarding the MIPS and the PowerPC datasets, we collect malware samples from the VirusShare[4] database. We collect the benign samples from Raspberry Pie II and architecture-specific compatible applications from the Linux repositories[5] and verify them by VirusTotal[6] to ensure benignity.

5.3.3.2 Feature Extraction

Algorithm 1 illustrates the feature extraction process from an ELF executable sample.

Files Parsing. The first step in parsing the collected binary files is to unpack and decompress the embedded application to obtain its assembly source code (ASM) files. We capture the binary CPU architecture from the ELF file metadata. Then, we employ objdump [85] to disassemble the files. We use a predefined matching pattern for extracting OpCodes from the gathered malware and benign unpacked samples based on the architecture type.

Generating OpCode Sequences. It is worth mentioning that the way these OpCode sequences are obtained and prepared can dramatically undermine embedded malware detection methods' accuracy and complexity. Accordingly, we give extra consideration to the pre-processing steps to ensure all information is captured. Each sample in the dataset includes a sequence of assembly OpCodes. Each sequence of OpCodes is then transmitted to a vector $Op_{s_1^n}$, where n is the total number of OpCodes in a sample Op_s. Then we reformulate and split the collected long Opcode sequence vector of each sample into vectors of 512 consequent OpCodes Op_split_i. The remaining OpCodes within a sequence are considered in the next row of the array Op_split_{i+1}. These reformatting and splitting processes continue until the last OpCode Op_n within an OpCode sequence, creating a matrix $Op\_split_1^N$ of 512 sequences, where N is the number of raws within the created matrix. Hence, all OpCodes within an application are taken into consideration as described in Equation 5.1; Equation 5.2:

$$Op_s = Op_1, Op_2, ...Op_n; \tag{5.1}$$

$$Op_s = Op\_split_1^N = \begin{bmatrix} Op_1 & Op_2 & Op_i & Op_{512} \\ Op_{513} & Op_{514} & Op_{split_l} & Op_{1024} \\ Op_j & Op_{j+1} & Op_{split_{j+l}} & Op_n \end{bmatrix} \tag{5.2}$$

---

[4]https://virusshare.com/

[5]https://rpmfind.net/linux/RPM/; https://pkgs.org/

[6]https://www.virustotal.com/

Algorithm 1: Feature Extraction for DeepWare

Input: ELF binary file

Result: Sentence embeddings of the OpCode sequence of the binary file, $\text{Emb}(\text{Op\_split})_1^N$

Begin

1   **Function** Files Parsing(*binaryfile*):

      $F \longleftarrow binaryfile$

      procedure(*Unpack and decompress*) F

      return $ASM file$;

        end procedure

2   **Function** disassemble(*ASM file*):

           $Arch \longleftarrow detect(Binary - Architecture)$;

3   **Function** extract OpCode_sequence(*ASM file*, *Arch*):

4   **while** (*not at the end of the file*){

               read current

5   **foreach** ($Op \in ASM file$)**do**

                 if ($Op \in Arch$){

                   $Op_s \longleftarrow op$;

             $Op_s = Op_1, Op_2, ... Op_n$; where n is the total number of OpCodes in an executable.

           End Function

        End Function

      return Op_s;

End Function

6   **Function** Generate OpCode_Sequences(Op_s):

      $i \longleftarrow 1$;

7   **foreach** ($Op_j \in Op_s$)**do**

        if ($j < i * 512$){

          $Op\_split_i \longleftarrow Op_j$;

        }

        else{

          $i \longleftarrow i + 1$;

          $Op\_split_i \longleftarrow Op_j$,

      $Op\_split = Op\_split_1^N$, N is the number of (512 OpCode) rows within the array. ;

      Op_split_i is the ith generated 512 length OpCode sequence vector. ;

      return Op_split;

End Function

8 **Function** Generate Sentence Embedding(Op_split):

9   **foreach** ($Op\_split_i \in Op\_split$)**do**

      Generate tokenized output vector, $T(Op\_split_i)$;

      Generate Sequence embedding, $\text{Emb}(Op\_split_i)$;

      return $\text{Emb}(Op\_split)_1^N$;

End Function

Sentence Embedding Extraction. Each sample is applied to a BERT tokenizer to transform an OpCode sequence into a numerical vector while maintaining the context of the OpCode. We pad the output sequences vector to ensure all samples have the same length. The tokenizer replaces each OpCode with its corresponding vector and creates a matrix $X \in R^{n*d}$ for the next operation, where n is the number of OpCodes with a sequence, and d is the dimension of output embedding.

Then we apply the tokenized data to a pre-trained BERT model embedder, specifically, the BERT classification model from [117] to capture the relations and semantics within OpCode sequences. The BERT model has 12 hidden layers as explained in [71], and we collect and concatenate the tensor output of all the hidden layers. Each collected matrix output dimension is then reduced by squeezing the batch dimension, creating an output two-dimensional tensors vector of size *(512*768)* for Op_split_i. The constructed matrix is then concatenated with the generated tensor for the rest of the sequence vectors of a sample $Op\_split_1^N$. Finally, the generated embedding matrix $Emb(Op\_split)_1^N$ is applied to the neural network model for training and classification.

5.3.4 MODELING

We develop an attention-based multi-head CNN-BiLSTM hybrid model that combines BiLSTM and CNN powers with attention mechanism benefits. The model design is depicted in Figure 5.2.

The CNN model uses convolutional layers and maximum pooling layers to learn from the raw data directly to secure higher-level features without the need for domain expertise to manually engineer input features. In contrast, BiLSTM models capture long-term dependencies between OpCodes in sequences, maintaining the overall sequence semantics. We use a customized attention layer based on Bahdanau et al. [21] after the BiLSTM layer to allow the classification layer to benefit from the overall context and to achieve better accuracy in the prediction. It is worth mentioning that we use a dropout layer with a 10% rate after each layer for regularization. To prevent overfitting, we use a Batch Normalization (BN) layer before the prediction layer. The BN layer is essential, as it decreases the internal covariance shift [41] that regularizes batches and accelerates training [302].

5.3.4.1 Input Layer

A two-dimensional input vector (512*768) created by the BERT embedding layer is fed in this layer. Then a two-dimensional output is passed to the CNN layer for further feature extraction and parameter tuning.

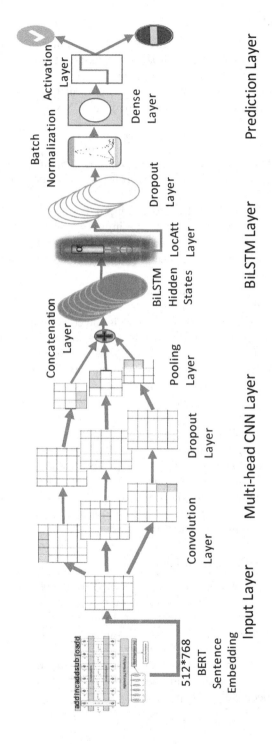

Figure 5.2: Our Proposed Hybrid Multi-Head CNN-BiLSTM-LocAtt Model Design.

5.3.4.2 Multi-head CNN Layer

We use multi-head CNN consisting of three parallel 1D CNNs heads to capture all the potential connections between adjacent OpCodes. We use filter sizes of 64 and kernel heads with varying sizes, precisely, a window size of 2, 3, and 4, to capture hidden associations within different adjacent OpCodes [122]. The number of each layer's parameters on a multi-head CNN is computed using Equation 5.3:

$$params = F_N * K_S * P_{PL} + bias \qquad (5.3)$$

where F_N represents the number of filters, K_S denotes the kernel window size, P_{PL} represents the last output vector from the previous layer, and $bias$ equals F_N.

The filter size is the number of times the input is processed, whereas the kernel size is the reading window size of the input sequence. Since we use a multi-head model, the model reads the input sequence using three different-sized kernels, enabling the model to read and understand the sequence data at three different resolutions.

The CNN head is composed of three stacked one-dimensional layers, including the convolution, dropout, and pooling layers. The convolution layer with Rectified Linear Units (ReLU) activation [119] is the core component that detects and extracts meaningful features of the input sample. The dropout layer is used to slow down the learning process and prevent overfitting. Finally, the pooling layer is mainly used to reduce the training time and diminish overfitting by decreasing the number of parameters, consolidating them to only essential elements. We use max-pooling layers with $pool\_size = 2$. Max pooling skims a window over its input and carries the max value in the pooling window, thus, downsampling each feature map independently, reducing dimensionality, and keeping the depth intact. Each of the CNN heads processes its input sequences window-by-window. A sequence of feature maps is then collected for each input sequence in consecutive order. A feature map FM_1^w is captured for each window input data, where w denotes the number of windows (in our case $w = 3$). Then, the window's output is concatenated with each other to create a sequence of feature maps.

5.3.4.3 BiLSTM with Attention Layer

The features map of each window created by concatenating each convolutional head's output feeds the BiLSTM layer. In this way, the BiLSTM works on the extracted feature maps in chronological order, from FM_1 to FM_w. The BiLSTM processes each window to collect the relevant information corresponding to the current window in its internal memory.

The BiLSTM layer with ReLU activation function captures forward and backward contextual information and has a powerful ability to extract significant information, specifically in capturing long-distance OpCodes dependencies. However, the BiLSTM layer processes each word sequentially [8]; thus, they cannot extract the local context information. Therefore, we use the Lo-cAtt layer after the BiLSTM layer to decide on the highly correlated features and diminish the number of learnable weights required for the prediction.

The BiLSTM generates an output vector from the sequence of the hidden states $(h_1, h_2, \ldots, h_n)$ for each input sequence, where n is the number of embedded representations of OpCodes in a sequence. We capture each hidden state and sequence of the BiLSTM encoder by setting $return\_state = True$ and $return\_sequence = True$ for the Keras LSTM function. We then concatenate the backward and forward hidden states of the BiLSTM layer by applying them to a concatenation layer. The vector h_j represents the concatenation of the forward and backward hidden state of the BiLSTM encoder as represented in Equation 5.4. The proposed customized attention mechanism is based on Bahdanau attention [21] that learns the network's weights and generates a context vector c_i for an output y_i using the weighted sum of the notes shown in Equation 5.5 [21]. The weights α_{ij} are calculated using the softmax function in Equation 5.6, where S_i is a BiLSTM hidden state for time i and e_{ij} is the attention score described by the function a that attempts to capture the association between input at j and output at i.

$$h_j = [\overrightarrow{h_f^T}, \overleftarrow{h_b^T}]^T \tag{5.4}$$

$$c_i = \sum_{j=1}^{n} \alpha_{ij} * h_j \tag{5.5}$$

$$\alpha_{ij} = \frac{\exp(e_{ij})}{\sum_{j=1}^{n} \exp(e_{ik})}; \quad e_{ij} = \alpha(s_{i-1}, h_j) \tag{5.6}$$

We set the LocalAtt layer with a window size of 30. Then a dropout layer and a BN layer are applied to regularize the activations and accelerate the model training.

5.3.4.4 Prediction Layer

The prediction layer is a dense layer with one unit, used to capture the attention layer's final features. Then we follow it with an output layer with a Sigmoid activation function that predicts the final results.

5.4 EXPERIMENTS

To evaluate our approach, we use the real-world ELF embedded applications format. This section presents the dataset, evaluation metrics used in our experiments, experimental environment, and the implementation details, and it reports the experiments performed and their results.

Table 5.1

Statistics of the Malware and Benign Samples Dataset.

ARM			
	Train	Test	Total
Malware	194	86	280
Benign	191	78	269
Total	385	164	**549**
Total Split OpCode Sequences	28,733	12,314	41,048
PowerPC			
	Train	Test	Total
Malware	496	193	689
Benign	121	67	188
Total	617	260	**877**
Total Split OpCode Sequences	20,060	8,600	28,660
MIPS			
	Train	Test	Total
Malware	273	127	400
Benign	251	98	349
Total	524	225	**749**
Total Split OpCode Sequences	34,965	14,785	49,750

5.4.1 DATASET AND EVALUATION METRICS

Dataset. As described in Section 5.3.3.1, we collect malware and benign samples from different architectures. We randomly divide our datasets into training sets (70%), validation sets (10%), and testing sets (20%). To generalize our experiments, we collect samples that include variants of the same sample from different architectures and different Linux and embedded Linux OS flavors. Table 5.1 lists the number of training and test samples used in each CPU architecture experiment. We merge the number of validation and testing samples to represent samples used for testing in the table. The total split OpCode sequences represent the number of OpCode sequences created from the splitting operation as described in Section 5.3.3.2 and in Equation 5.1 and Equation 5.2.

Evaluation Metrics. This work is presented as a binary class classification problem to detect malware samples. Classification metrics such as Receiver Operating Characteristics (ROC) , False Positives (FP), False Negatives (FN), True Positives (TP), and True Negatives (TN) are used to show the accuracy and correctness of the proposed model. The precision metric confirms how many of the identified malware traces are correct, and the recall metric depicts how many of the malicious samples the model detects. Furthermore, the F1-score metric can unite both precision and recall metrics to show the average. The F1-measure relationship and the association between the correctly and incorrectly identified cases with F1-score are shown in Equation 4.3.

Table 5.2

The Hyper-Parameter Specification for All the Model Layers.

Parameter	Value
Conv. Heads	3
Conv. filter	64
Kernel window sizes	2,3,4
MaxPolling size	2
BiLSTM units	8
Droupout	0.1
Attention window size	30
Output Dimension units	1
Learning rate	0.0001
Epochs	100 (early stop is set)
Batch size	20

5.4.2 IMPLEMENTATION DETAILS

We use BERT to extract the sequence embedding and TensorFlow [92] to implement the model, and Keras [133] on top of TensorFlow to aid the implementation and experimentation. We test our model on a server with Core-i7 (3.6 GHz) processor, 16 GB RAM, and a computer with Core-i5 (2.3 GHz) processor, 8 GB RAM. Our experimental environment is as follows: Ubuntu 18.04.5, Python 3.6.7, CUDA 11.0, Transformers 3.0.2, TensorFlow 2.2.0, and Keras 2.4.3.

We build a functional CNN-BiLSTM model using the Keras deep learning library. The model expects a three-dimensional input with [batch_size, sequence_length, features_embedding_dimension]. We set the batch size to 20, and the epoch is 100 while setting EarlyStopping monitoring the validation loss with the patience of 15 epochs. The output of the model is one vector containing the probability of a given window indicating malware or benign.

The training process is carried out in a supervised manner, using backpropagation to propagate gradients from the final dense layer to the top convolutional layer. Since our concern is to detect malware, we classify our problem as a binary classification task; thus, the binary cross-entropy loss function is used. We use Adam [308] optimizer with a learning rate of 0.0001. Table 5.2 depicts the hyper-parameter specifications.

5.4.3 RESULTS DISCUSSION

To examine and evaluate our proposed model's competency for embedded device malware detection, we conduct two different approaches to obtain comparable results. The primary experiment is on the ARM-based dataset described in Section 5.4.3.1, followed by the experiments in Section 5.4.3.2 that

Table 5.3

Comparison Between the Proposed Model and the State-of-the-Art OpCode Based Malware Detection Approaches.

Model	Accuracy%	F1-Scores%	Precision%	Recall%	Year
DeepWare	99.39	99.40	99.40	99.40	2020
[19]	99.68	98.48	98.59	98.37	2019
[103]	94	-	-	-	2018

include the results of our model when examined on the three different CPU architectures and comparison of the deep learning model with the baseline classification algorithms. The experiments are performed using 10-fold cross-validation, while shuffling and splitting the datasets into training, validation, and test sets.

5.4.3.1 Preliminary ARM-Based Results

In our preliminary experiments, we cover only the ARM-based embedded dataset collection to validate our proposed model's performance. We choose the ARM dataset to compare the results with the state-of-the-art methods [19, 103] because they are realized and tested on ARM datasets only. We train our model with 70% of the data and validate it with the rest of the malware and benign samples. We use 10-fold cross-validation and achieve an average DR of 99.2%, FRR of 1.56%, and F1-score of 99.4%.

We conduct a set of experiments to select the optimal model parameters. Figure 5.3 illustrates some of these experiments to choose the number of CNN heads, batch size, and the number of epochs. In particular, Figure 5.3(a) presents a comparison between the number of heads in the multi-head CNN layer, showing that three heads give the best performance with an F1-score of 0.9946%. Figure 5.3(b) illustrates the batch size selection experiment, where a batch size of 20 produces the best results. Figure 5.3(c) shows that the lowest validation and training losses are achieved after 20 epochs. We set the ModelCheckpoint and the EarlyStopping parameters to monitor the validation loss to stop the training and save the best-performing trained model.

We compare our performance results with the recent malware detection studies in the literature that depends on OpCode sequence features for IoT-based malware detection using deep learning methods. Azmoodeh et al. [19] combined the information gain technique and the graph representation of selected OpCodes to create features. These features were then applied to Eigenspace and deep convolutional networks to classify samples. On the other hand, HaddadPajouh et al. [103] expressed that the two-layer LSTM network achieved the best accuracy to classify OpCodes samples. Table 5.3 displays

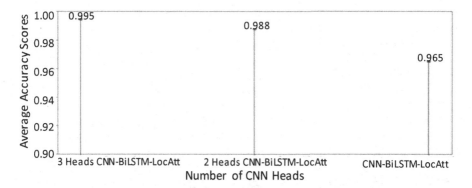

(a) (a) Tuning The Number Of CNN Heads Parameter.

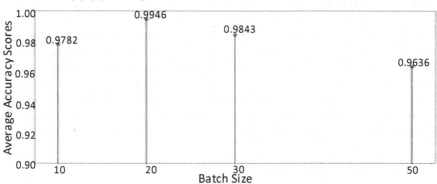

(b) (b) Tuning The Batch Size Parameters.

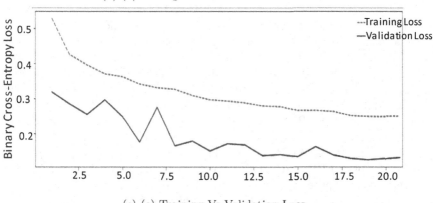

(c) (c) Training Vs Validation Loss.

Figure 5.3: ARM Dataset Different Parameter Setting Results. (a) Results for Different CNN Head Numbers. (b) Results for Different Batch Sizes. (c) Training and Validation Loss Throughout the Number of Epochs.

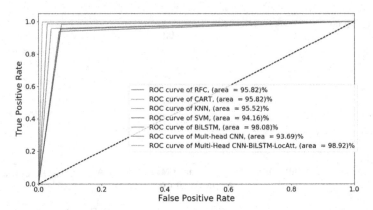

Figure 5.4: ARM ROC Curve and AUC for Baseline Models in Comparison with Our Model.

the accuracy and F-measure scores comparing our model with recently presented OpCode-based malware detection models. By comparing our results with the ones from the state-of-the-art approaches, we slightly improve the detection accuracy for ARM-based malware detection. However, due to the lack of similar work in the literature that considers other CPU architectures, we can not present a comprehensive comparison. Accordingly, we validate our results with several baseline models, to be discussed in Section 5.4.3.2.

5.4.3.2 Baseline Algorithms Comparison

In this section, we compare our results with baseline classifiers, namely Random Forest Classifier (RFC), Decision Tree (CART), k-Nearest Neighbor (KNN), Support Vector Machine (SVM), BiLSTM, and Multi-head CNN. We first compare our model and the baseline models using the ARM dataset. We then perform a comprehensive performance evaluation of our malware detection solution using all our CPU architecture datasets utilizing standards metrics, comparing our results to the baseline classification techniques. The ROC is a 2D space curve that shows the True-Positive Rate (TPR) and the False-Positive Rate (FPR) for each classifier. The Area Under the Curve (AUC) is the area under the ROC curve and can evaluate a binary classifier's performance. Figure 5.4 shows the comparison between the ROC and AUC for each classifier for the ARM dataset. The results indicate that our approach produces a lower FPR and a higher DR than all of the baseline classifiers.

Figure 5.5 compares the classification F1-score of our model compared to the baseline classifiers for all of the CPU architectures in our dataset. Our proposed solution achieves F1-scores of 99.4%, 97.1%, and 94.2% for the ARM, PowerPC, and MIPS datasets, respectively. The results prove that our proposed approach is better than all the baseline classifiers with consistently high accuracy for all architectures.

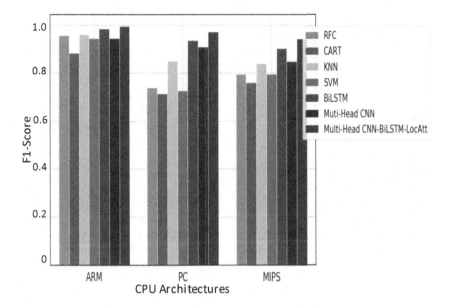

Figure 5.5: A Comparison Between Our Multi-Head CNN-BiLSTM-LocAtt Model and the Baseline Models. The Comparison Is Conducted on the Three CPU Architecture Datasets.

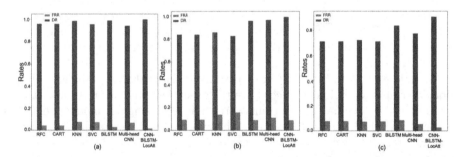

Figure 5.6: Detection Rate (DR) and False Rejection Rate (FRR) for All the Baseline Models Against Our Model. (a) Results of ARM dataset. (b) Results of PowerPC dataset. (c) Results of MIPS dataset.

Figure 5.6 illustrates a comparison of the DR and the FRR achieved by our solution against the baseline techniques. Figure 5.6(a) denotes the results for the ARM dataset experiments, and our approach achieves an almost perfect DR and a very low FRR of 1.26%. Figure 5.6(b) shows the DR and FRR results for the PowerPC dataset. The results indicate that our model produces the highest DR of 98.99% and FRR of 8.7%. We observe high FRR results in this experiment, which is mainly due to the highly imbalanced data samples

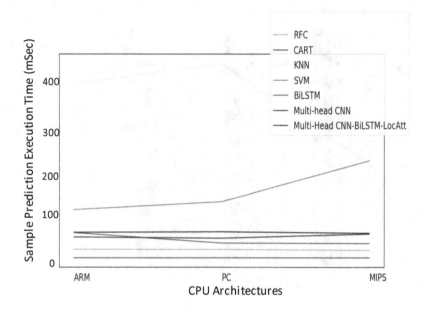

Figure 5.7: Comparison of Prediction Time on the Three CPU Architecture Datasets: All Baseline Models vs. Our Model.

of the PowerPC dataset, as shown in Table 5.1. Figure 5.6(c) presents the experimental results on the MIPS dataset where our model achieves over 90% DR and less than 3% FRR.

To validate that our model can be used as a near real-time detection solution, we run an experiment to compare the execution time taken to predict if an unknown binary file sample is malware or benign using our model and the baseline models. Figure 5.7 illustrates the experimental results showing that our deep-learning model is almost stable across all CPU architectures with an average of 62 ms for prediction time. CART and RFC classify samples in the least amount of time, which is less than 10 ms. Our model performs similarly to BiLSTM and slightly higher than multi-head CNN in the ARM experiments. However, it classifies PowerPC and MIPS samples slower than both. Based on our experimental results, SVM and KNN are consistently the slowest classifying models.

Regarding the training elapsed time, Table 5.4 demonstrates the training time for our approach compared to the baseline models. The training of our model spends around 15 mins, with several seconds less than the BiLSTM model. SVM is the slowest model to train. On the other hand, KNN and CART are the fastest models to train, with fewer than 2 mins.

Table 5.4

Training Execution Time of Our Multi-Head CNN-BiLSTM-LocAtt Model in Comparison to the Baseline Models.

Model	Training Elapsed Time
RFC	00:01:42.587812000
CART	00:01:20.742626000
KNN	00:01:02.437473000
SVM	11:12:19.223035000
BiLSTM	00:15:35.452159000
Multi-head CNN	00:11:45.130163
Our model	**00:15:30.978338000**

Note: The Time Format is HH:MM:Sec.mSec.

5.5 SUMMARY

This chapter proposes a novel real time embedded systems' malware detection solution named DeepWare that uses CPU architectures, OpCode sequences, and DL mechanisms to detect malware. We extract BERT sentence embedding from the binary file OpCode sequences to maintain the sequence's semantic meaning and context. Then we apply the generated sentence embedding to a hybrid multi-head CNN-BiLSTM model with local attention to extracting the inherent and correlated features within a sequence. The multi-head CNN extracts the features of an OpCode sequence on an entirely independent basis. Afterward, the BiLSTM extracts the long-term forward and backward relationship between OpCodes, passing the information to a Location Attention Layer (LocAtt) that determines the highly correlated features. Finally, a prediction layer is used to determine whether a sample is benign or malware. We implement and evaluate DeepWare on three different embedded CPU architectures: ARM, MIPS, and PowerPC. Our extensive evaluation results demonstrate DeepWare's potential to efficiently detect embedded systems' malware, specifically embedded Linux malware variants targeting the three different CPU architectures. We prove through statistical tests that DeepWare using hybrid multi-head CNN-BiLSTM with attention architecture outperforms the traditional classification models. Moreover, we compare the results of the ARM dataset experiments with the other state-of-the-art OpCodes-based malware detection models. The experimental results show an encouraging improvement in the detection accuracy for ARM-based malware.

6 Conclusion and Future Directions

In this chapter, we summarize the contributions of this book and discuss some future research directions regarding IoT security.

6.1 SUMMARY

IoT brings the physical and digital worlds together in the sense that IoT is creating a seamless integration of physical things into communication networks. Offering the capability of integrating both digital and physical entities, IoT becomes a vital model that enables a whole new class of applications and services. Security is one of the most challenging issues that need to be addressed before these IoT applications and services can be fully embraced. In this book, we address several research issues regarding IoT infrastructure security. We study the significant research efforts throughout 2013-2020 that address IoT security and privacy issues to understand the problem and current gaps. We tackle IoT threat detection and device identification problems. We design a security framework that integrates device identification, access control, and intrusion detection for IoT infrastructures. We suggest a novel behavioral fingerprinting technique for IoT devices. We propose using the fingerprinting method with ML to identify IoT devices connecting to the network and detect any malicious communication. We offer an application layer malware detection solution that detects malicious binary files targeting IoT and embedded systems. We also provide an implementation of our approaches for BIN-IoT and DeepWare. In particular, we summarize our main research contributions in the following:

- Survey on IoT Security Mechanisms and Challenges: We present a detailed survey that includes more than 100 information security-related works over the period of 2013-2020, while a number of them are specifically for securing the IoT with a focus on their diversity. The security and privacy issues in each of the three layers of the IoT system architecture have been studied independently under the theme of IoT device security, cloud/edge security, and application security. However, such independently developed security solutions miss the point that IoT services and applications are delivered by collecting data from the physical layer, processing them at the devices/edges/cloud, and then being accessed or consumed by users via applications and services. To the best of our knowledge, none of the previous surveys looks at the security solutions at the information processing layer while considering the interactions with the physical and application layers. To the best of our knowledge, this is the first survey that attempts to include all security aspects and security challenges facing the information layer of IoT solutions including outsourcing techniques for partial computations on edge or cloud, while presenting case studies to map security challenges and requirements in real IoT scenarios.

- Security Framework Design: We present a self-learning real-time security framework design, named Behavioral Network Traffic Identification and Anomaly Detection for the Internet of Things (BIN-IoT). BIN-IoT can aid network administrators to control access to IoT devices by identifying connecting devices and providing each identified device with its predefined privileges. Each device has a predefined set of privileges including a communication zone. Any behavioral alteration for any of the communicating IoT devices from the behavioral baseline can be detected by a continuous random moving window fingerprinting of the devices. Hence, the identified compromised devices are quarantined for further monitoring.

- Network traffic behavioral fingerprinting: One of the challenges of big data produced from IoT devices is managing high-dimensional data. Accordingly, feature selection and extraction are key to enhancing IoT device identification and malicious traffic and malware detection. Fingerprinting is one of the famous techniques to select and extract unique behavioral features of IoT devices. We propose a novel behavioral fingerprinting technique for IoT infrastructure that captures the behavior patterns of IP-enabled IoT devices' network traffic. This behavior-based fingerprint is applied to either the identification module (BI-IoT) or the anomaly detection module (BND-IoT) that uses ML solutions to identify IoT devices or to detect malicious traffic, respectively. Integrating these modules allows the system to provide adequate access control and identify compromised IoT or rogue devices connected to the network. The presented fingerprint study shows that it can identify device types accurately with only a small sequence of packets.

- IoT Device Identification: We propose an IoT device and device type identification system, named Behavioral IoT Network Traffic Identification (BI-IoT) that uses our novel fingerprinting technique with supervised ML on the network traffic extracted from IP-enabled IoT devices to recognize multiple known IoT devices and device types (from white list). The presented fingerprint study shows that with a sequence of a limited number of packets, it can recognize a device type accurately. Moreover, unknown or rogue devices connecting to the network can be identified. Our proposed approach can identify individual devices of the same type (same vendor and model), which is an essential feature to manage the network and the access control solutions for each device, not only the device type.

- IoT Network Traffic Novelty Detection: In the absence of any prior expert knowledge on anomalous data, we propose Behavioral Novelty Detection for the IoT Network Traffic (BND-IoT). BND-IoT is a real-time IoT network traffic anomaly detection system that can detect anomalous traffic from unknown, unseen attacks, and malware traffic when the network model is trained with normal traffic only. BND-IoT uses our proposed novel behavioral fingerprinting technique that captures the behavior patterns of IP-enabled IoT devices' network traffic. This behavior-based fingerprint can be used by the novelty detection solution to catch malicious traffic, thus allowing the system to identify compromised IoT or rogue devices connected to the network.

- Embedded Devices Malware Detection: We propose DeepWare, a real-time cross-architecture malware detection solution for embedded devices. DeepWare can run on either network devices or edge devices to detect malware. We validate our method by collecting OpCodes datasets and running all the experiments on three of the most commonly used embedded microcontroller architectures: ARM, PowerPC, and MIPS. DeepWare combines BERT sentence embedding and CNN together with BiLSTM DL techniques for embedded malware detection. We use the OpCode BERT embedding sequence to extract the fine-grained inherent connection and the semantic behaviors between embedded programs. These extracted features are then fed into a DL network based on CNN-BiLSTM, which considers the hidden malicious program behaviors to detect malware.

- IoT Malware Identification DL Network: We construct a malware identification network of multi-headed CNN and BiLSTM networks for mining correlation and context information between different OpCodes, and a LocAtt mechanism to re-adjust the importance (weights) of the correlated features from OpCodes and their sequences to improve the DL ability to detect malware. We name the DL model multi-head CNN-BiLSTM-LocAtt. The LocAtt is seamlessly compounded with CNN-BiLSTM since the LocAtt weights for feature readjustment are calculated from both features and the hidden states of the CNN-BiLSTM. To the best of our knowledge, this is the first time a LocAtt mechanism is used for embedded devices' OpCode sequence malware detection.

- Datasets Collection: Based on our observations, we believe that there is a need to collect and extract normalized OpCode datasets of IoT malware and benign applications for the three most common microcontroller architectures currently available. Accordingly, we publish a normalized OpCode dataset of IoT malware and goodware applications for ARM, PC, and MIPS CPU architectures. We believe this dataset can contribute significantly to the development of the emerging embedded systems malware detection research area.

6.2 FUTURE DIRECTIONS

In recent years, although active research efforts have been dedicated to securing IoT, there are still a number of open issues that need to be tackled. This book addresses a number of issues facing the data and network security of IoT systems. However, there are many opportunities to extend this work for securing and privacy preservation of IoT systems. In this section, we lay out a research agenda by proposing several future research directions.

Comprehensive Security. Several promising opportunities [20, 73, 112, 142, 206, 307] have emerged with the evolution of IoT and cloud computing, facing IoT security challenges [184, 237, 257, 310]. However, a complete solution has not yet been implemented, as most of the available solutions target only certain security requirements. To the best of our knowledge, there is still no methodology that develops security attack-free, conditional anonymous authentication and fine-grained access control protocol to be used by the various IoT use case scenarios. This framework should also outsource part of the heavy computations in either uploading or downloading data, to a computational powerful device or the cloud. Existing techniques can be integrated to accomplish an all-inclusive security solution. However, the integration is not simple, due to the possibility of technique interference, usability, and efficiency problems. We believe that comprehensive security solutions are needed for usable and efficient IoT data sharing.

Non-Repudiation. In any system involving user interactions, non-repudiation is needed to prevent the data owner (sender) from the denial of previous data uploads. Most of the work in IoT security, however, neglected the importance of non-repudiation. IoT systems concentrate more on the privacy of users and devices. However, non-repudiation can still be provided, while preserving user privacy. This can be achieved by conditional anonymity, as the sender still needs to be anonymous, and for security or emergency reasons, certain authorities need to find out the real identity of a user. Conditional anonymity can be achieved by a number of techniques such as group signatures. However, providing a conditional anonymity feature on limited-resourced IoT devices is a challenging task, and extensive research is needed in this direction.

Scalability. IoT devices interact in different patterns with different entities. The number of users and devices communicating together within an IoT ecosystem is growing widely with the growth in technologies. Accordingly, the efficiency and scalability of systems need to be ensured by not only storing secured data on the cloud, but also by delegating the encryption and decryption in a secured manner to a cloudlet or cloud. Outsourcing the revocation management and access list management to the cloud (proxy) is also needed, to reduce the computational load of both sender and receiver in different perspectives. Additionally, the size of the encrypted data (sent or received) should be minimized to reduce communication overhead, which can be achieved by using constant size Attribute-Based Encryption (ABE) encryption techniques.

Moreover, the current IoT anomaly detection solutions should be scalable enough to be applied to similar Industrial systems such as Supervisory Control and Data Acquisition (SCADA) systems. Both cloud computing and SCADA systems are exposed to the Internet for operation. Hence, they require a lightweight and scalable NADS for detecting abnormal activities within their networks. Special consideration should be taken to deploy NADS solutions in these large-scale environments, as they usually consist of a large number of nodes and require minimum latency in their usual operations.

Revocation. As the user/device may have limited subscription periods or the device has been hacked/attacked or stolen, the communicating party needs to find out whether a user/device is revoked. The revoked entity authentication and access to data should be disabled. Efficient revocation is very challenging and it is especially important for a large-scale network. Scalable key management revocation in IoT systems is an important aspect. It should ensure both backward and forward secrecy. Newly joined users to a group, should not be able to interpret data encrypted, before their joining time. Revoked users that previously had a key, should not be able to interpret future encrypted data using their previous revoked keys. Finding the proper and efficient way to revoke attributes and/or users is still an issue. The existing schemes are not flexible nor efficient enough. For instance, some existing techniques are based on users interacting online with the authority [28]. Some techniques do not allow revocation instantly [297]. More research is therefore needed to enhance the user and attribute revocation and management systems. Outsourcing the management to a semi-trusted cloud can add flexibility and scalability to infrastructure management systems (issuing and revocation of keys and attributes).

Interoperability. Interoperability in IoT is needed, knowing that there are some legacy proprietary hardware and software deployed systems. Governmental efforts to create standards for IoT interoperability and backward compatibility should be taken into consideration during all phases of IoT systems implementation and fabrication. Moreover, governmental and non-governmental entities should provide a way to set universal privacy policies and find a way to impose them. This is needed to allow the interoperability of different systems while maintaining user privacy.

Trust Management. Trust management is needed to establish trust across different IoT systems and domains. In such a large-scale IoT network, it is a big challenge to build trust between different domains and a large number of limited resources devices [54]. The number of IoT consumers can be huge and dynamic. Accordingly, the trust management system must be adaptive and scalable. Moreover, the accuracy of trust systems is crucial. To improve the trust result accuracy, different techniques can be integrated such as reputation and recommendation techniques. However, most of the available trust management systems cannot be easily integrated with each other [198]. Accordingly, more research is needed to provide techniques to efficiently integrate various trust feedbacks. This should be done while protecting users' privacy. Furthermore, the response time of trust management systems is of great importance. The longer the response time, the lower the number of inquiries that the system can handle. Systems with minimal response time are urgently needed for IoT systems.

Latency Constraint. It is important to note that not only the information in IoT should be stored and processed in a timely manner, but also services should be secured with minimal added latency. It is not recommended to rely on the cloud for applications that require fast processing and minimal delays [239]. Edge computing and federated-machine learning [293] are some of the promising ways that can ensure the rapid delivery of IoT cloud-based services as well as scalability and privacy-policy enforcement. More research efforts are needed in finding efficient ways to use edge computing with IoT and the cloud to combat the security challenges in IoT systems.

Datasets and Testbed Configurations. One of the current challenges that face this work and other similar research efforts is the absence of publicly available datasets that contain the recent host and attack vectors targeting IoT devices. The creation of such a dataset demands configuring a suitable network environment while capturing host and network activities concurrently to evaluate both host and network-based IDSs. Novel attack types should be explored further in the datasets. These datasets can help researchers investigate the integration of host and network ID features to detect IoT network traffic intrusions more efficiently with a lower False Positive Rate (FPR).

Hybrid Malware Detection. A robust and practical solution is required to analyze IoT applications and advise users if they hold obfuscated code that might cause malicious activities. Obfuscation methods hide malware and try to avoid getting detected by anti-malware solutions. Moreover, polymorphic techniques, such as inserting dead code into the original application code, can be used to evade malware detection. As simple static analysis usually can not detect obfuscated code, at this stage, investigating the inclusion of more static features such as API call sequence, and the integration of static and dynamic analysis are required to enhance the malicious code's recognition. The investigation should be conducting more experiments and studies to detect malware in a real-time environment or on a live system to obtain practical results.

References

1. Humberto Abdelnur, Olivier Festor, Jérôme François, and Radu State. Machine Learning Techniques for Passive Network Inventory. *IEEE Transactions on Network and Service Management - TNSM*, 7:244–257, 2010.
2. Rahul Agarwal. NLP Learning Series (Part 3): Attention, CNN and what not for Text Classification, 2019. URL: https://towardsdatascience.com/.
3. Charu C. Aggarwal, Naveen Ashish, and Amit Sheth. The Internet of Things: A Survey from the Data-Centric Perspective. In *Managing and Mining Sensor Data*, pages 383–428. Springer US, 2013.
4. Emmanuel Ahene, Junfeng Dai, Hao Feng, and Fagen Li. A certificateless signcryption with proxy re-encryption for practical access control in cloud-based reliable smart grid. *Telecommunication Systems*, 70:491–510, 2018.
5. L. Akoglu and H. Tong. Graph based anomaly detection and description: A survey. *Data Mining and Knowledge Discovery*, 29:626–688, 2015.
6. K. Eykholt et Al. Robust physical-world attacks on machine learning models. *IEEE Conference on Computer Vision and Pattern Recognition -CVF*, pages 1625–1634, 2018.
7. Ammar Alazab, Michael Hobbs, Jemal Abawajy, and Ansam Khraisat. Using response action with Intelligent Intrusion detection and prevention System against web application malware. *Information Management and Computer Security - IMCS*, 22:431–449, 2014.
8. Mamoun Alazab. Profiling and classifying the behavior of malicious codes. *Journal of Systems and Software*, 100:91–102, 2015.
9. Gergely Alpár, Lejla Batina, Lynn Batten, Veelasha Moonsamy, Anna Krasnova, Antoine Guellier, and Iynkaran Natgunanathan. New directions in IoT privacy using attribute-based authentication. *ACM International Conference on Computing Frontiers - CF*, pages 461–466, 2016.
10. Nishadh Aluthge. *IoT device fingerprinting with sequence-based features*. PhD thesis, HELSINKI, 2017.
11. Ruhul Amin, Neeraj Kumar, G. P. Biswas, R. Iqbal, and Victor Chang. A light weight authentication protocol for IoT-enabled devices in distributed Cloud Computing environment. *Future Generation Computer Systems*, 78:1005–1019, 2018.
12. G S G N Anjaneyulu, P Vasudeva Reddy, U M Reddy, and Asso Professor. Secured Digital Signature Scheme using Polynomials over Non-Commutative Division Semirings. Technical Report 8, 2008.
13. Ionut Arghire. Mysterious Hajime Botnet Grows to 300,000 IoT Devices: Kaspersky, 2017. URL: http://www.securityweek.com/mysterious-hajime-botnet-grows-300000-iot-devices-kaspersky.
14. Arm MBED. Mbed. URL: https://www.mbed.com/en/.
15. Michael Armbrust, Armando Fox, Rean Griffith, Anthony D. Joseph, Randy H Katz, Andrew Konwinski, Gunho Lee, David A. Patterson, and Ariel Rabkin. *Above the Clouds: A Berkeley View of Cloud Computing*. PhD thesis, Armbrust2009, 2009.

16. Muhammad Asim and Tanya Ignatenko. Attribute-based encryption with encryption and decryption outsourcing. *Australian Information Security Management Conference*, pages 21–28, 2014.

17. Giuseppe Ateniese, Kevin Fu, Matthew Green, and Susan Hohenberger. Improved proxy re-encryption schemes with applications to secure distributed storage. *ACM Transactions on Information and System Security*, 9:1–30, 2006.

18. Luigi Atzori, Antonio Iera, and Giacomo Morabito. The Internet of Things: A survey. *Computer Networks*, 54:2787–2805, 2010.

19. Amin Azmoodeh, Ali Dehghantanha, and Kim Kwang Raymond Choo. Robust Malware Detection for Internet of (Battlefield) Things Devices Using Deep Eigenspace Learning. *IEEE Transactions on Sustainable Computing*, 4:88–95, 2019.

20. Sachin Babar, Antonietta Stango, Neeli Prasad, Jaydip Sen, and Ramjee Prasad. Proposed embedded security framework for Internet of Things (IoT). *International Conference on Wireless Communication, Vehicular Technology, Information Theory and Aerospace and Electronic Systems Technology, Wireless - VITAE*, 2011.

21. Dzmitry Bahdanau, Kyung Hyun Cho, and Yoshua Bengio. Neural machine translation by jointly learning to align and translate. *International Conference on Learning Representations - ICLR*, pages 1–15, 2015.

22. Balbix. IoT Security Challenges and Problems, 2024. URL: https://www.balbix.com/insights/addressing-iot-security-challenges/.

23. Mrinmoy Barua, Xiaohui Liang, Rongxing Lu, and Xuemin Shen. PEACE: An Efficient and Secure Patient-centric Access Control Scheme for eHealth Care System. In *International Workshop on Security in Computers, Networking and Communications*, pages 987–992, 2011.

24. Zahra Bazrafshan, Hashem Hashemi, Seyed Mehdi Hazrati Fard, and Ali Hamzeh. A survey on heuristic malware detection techniques. *Conference on Information and Knowledge Technology*, pages 113–120, 2013.

25. Sana Belguith, Nesrine Kaaniche, Maryline Laurent, Abderrazak Jemai, and Rabah Attia. Constant-size threshold attribute based signcryption for cloud applications. In *International Joint Conference on e-Business and Telecommunications - ICETE*, pages 1–15, 2017.

26. C. Bellinger, Sh. Sharma, and N. Japkowicz. One-class versus binary classification: Which and when? *International Conference on Machine Learning and Applications - ICMLA*, pages 102–106, 2012.

27. Elisa Bertino, Kim-Kwang Raymond Choo, Dimitrios Georgakopolous, and Surya Nepal. Internet of Things (IoT): Smart and Secure Service Delivery. *ACM Transactions on Internet Technology*, 16:1–7, 2016.

28. John Bethencourt, Amit Sahai, and Brent Waters. Ciphertext-Policy Attribute-Based Encryption. *IEEE Symposium on Security and Privacy - SP*, pages 321–334, 2007.

29. Bruhadeshwar Bezawada, Maalvika Bachani, Jordan Peterson, Hossein Shirazi, Indrakshi Ray, and Indrajit Ray. IoTSense: Behavioral Fingerprinting of IoT Devices. *arXiv:1804.03852v1*, 2018.

30. Vitor Hugo Bezerra, Victor Guilherme Turrisi da Costa, Sylvio Barbon Junior, Rodrigo Sanches Miani, and Bruno Bogaz Zarpelão. IoTDS: A one-class

classification approach to detect botnets in internet of things devices. *Sensors*, 19:1–20, 2019.

31. Kemal Bicakci and Nazife Baykal. Server Assisted Signatures Revisited. *Topics in Cryptology - CT-RSA*, pages 143–156, 2004.

32. John Biggs. Hackers release source code for a powerful DDoS app, Mirai TechCrunch, 2016. URL: `https://techcrunch.com/2016/10/10/hackers-release-source-code-for-a-powerful-ddos-app-called-mirai/`.

33. David Bisson. New 'Kaiji' Linux Malware Targeting IoT Devices, 2020. URL: `https://securityintelligence.com/news/new-kaiji-linux-malware-targeting-iot-devices/`.

34. Matt Blaze, Gerrit Bleumer, and Martin Strauss. Divertible protocols and atomic proxy cryptography. *International Conference on the Theory and Applications of Cryptographic Techniques - EUROCRYPT*, pages 127–144, 1998.

35. Sören Bleikertz, Anil Kurmus, Zoltán A. Nagy, and Matthias Schunter. Secure cloud maintenance: Protecting workloads against insider attacks. In *ACM Symposium on Information, Computer and Communications Security - ASIACCS*, pages 83–84, 2012.

36. Sara Boddy and Justin Shattuck. The hunt for IoT: The rise of the thingbots. *F5 Labs Threat Analysis Report*, 3:1–27, 2017.

37. Dan Boneh and Matthew Franklin. Identity-Based Encryption from the Weil Pairing. *Annual International Cryptology Conference - CRYPTO*, 2139 LNCS:213–229, 2001.

38. Anna L. Buczak and Erhan Guven. A Survey of Data Mining and Machine Learning Methods for Cyber Security Intrusion Detection. *IEEE Communications Surveys and Tutorials - COMST*, 18:1153–1176, 2016.

39. D. P. Robinson C. You and R. Vidal. Provable selfrepresentation based outlier detection in a union of subspaces. *IEEE Conference on Computer Vision and Pattern Recognition - CVPR*, pages 1–10, 2017.

40. Jan Camenisch and Anna Lysyanskaya. An Efficient System for Nontransferable Anonymous Credentials with Optional Anonymity Revocation. In Birgit Pfitzmann, editor, *Advances in Cryptology - EUROCRYPT*, pages 93–118, Berlin, Heidelberg, 2001. Springer Berlin Heidelberg.

41. Mikel Canizo, Isaac Triguero, Angel Conde, and Enrique Onieva. Multi-head CNN–RNN for multi-time series anomaly detection: An industrial case study. *Neurocomputing*, 363:246–260, 2019.

42. Canonical. Ubuntu Core | Ubuntu, 2019. URL: `https://ubuntu.com/core`.

43. Larry Cashdollar. As We Warned, Iran Strikes Back with new Silex Malware Bricking IoT Devices, 2019. URL: `https://blog.tmcnet.com/blog/rich-tehrani/security/as-we-warned-iran-strikes-back-with-new-silex-malware-bricking-iot-devices.html`.

44. Ceshine Lee. Feature Importance Measures for Tree Models — Part I, 2017. URL: `https://medium.com/the-artificial-impostor/feature-importance-measures-for-tree-models-part-i-47f187c1a2c3`.

45. Varun Chandola and Arindam Banerjee. Anomaly Detection: A Survey. *ACM Computing Surveys - CSUR*, 41:1–58, 2009.

46. Nishanth Chandran, Jens Groth, and Amit Sahai. Ring Signatures of Sublinear Size Without Random Oracles. *Automata, Languages and Programming*, pages 423–434, 2007.

47. Suresh N. Chari and Pau Chen Cheng. Bluebox: A policy-driven, host-based intrusion detection system. *ACM Transactions on Information and System Security - TISSEC*, 6:173–200, 5 2003.
48. David Chaum. Blind Signatures for Untraceable Payments. *Advances in Cryptology*, pages 199–203, 1982.
49. David Chaum and Eugène van Heyst. Group Signatures. In Donald W Davies, editor, *Advances in Cryptology - EUROCRYPT*, pages 257–265, Berlin, Heidelberg, 1991. Springer Berlin Heidelberg.
50. Cheng Chen, Jie Chen, Hoon Wei Lim, Zhenfeng Zhang, and Dengguo Feng. Combined Public-Key Schemes: The Case of ABE and ABS. *International Conference on Provable Security - Provable Security*, 7496 LNCS:53–69, 2012.
51. Fei Chen, Yiliang Han, Di Jiang, Xiaoce Li, Xiaoyuan Yang, Di Jiang Xiaoce Li Fei Chen, Yiliang Han, and Xiaoyuan Yang. Outsourcing the Unsigncryption of Compact Attribute-Based Signcryption for General Circuits. *Social Computing*, pages 533–545, 2016.
52. Xiaofeng Chen, Jin Li, and Willy Susilo. Efficient Fair Conditional Payments for Outsourcing Computations. *IEEE Transactions on Information Forensics and Security - XPL*, 7:1687–1694, 2012.
53. Xin Ma Chen, Shize Guo, Haiying Li, Zhisong Pan, Junyang Qiu, and Y. Ding and Feiqiong. How to Make Attention Mechanisms More Practical in Malware Classification. *IEEE Access*, 7:155270–155280, 2019.
54. Wen-long Chin, Wan Li, and Hsiao-hwa Chen. 08067688. *IEEE Communications Magazine*, (October):70–75, 2017.
55. Sherman S. M. Chow, Joseph K. Liu, Victor K. Wei, and Tsz Hon Yuen. Ring Signatures without Random Oracles. *ACM Symp. Inform. Computer, Communication Security*, pages 297–302, 2006.
56. Hsin Yu Chuang and Sheng De Wang. Machine Learning Based Hybrid Behavior Models for Android Malware Analysis. *IEEE International Conference on Software Quality, Reliability and Security - QRS*, pages 201–206, 2015.
57. Kai-Min Chung, Yael Tauman Kalai, Feng-Hao Liu, and Ran Raz. Memory Delegation. *Advances in Cryptology - CRYPTO*, pages 151–168, 2011.
58. Catalin Cimpanu. Hajime IoT Worm Considerably More Sophisticated than Mirai, 2016. URL: http://news.softpedia.com/news/hajime-iot-worm-considerably-more-sophisticated-than-mirai-509423.shtml.
59. Cédric Clastres. Smart grids: Another step towards competition, energy security and climate change objectives. *Energy Policy*, 39:5399–5408, 2011.
60. Wave Computing. Wave Computing Launches the MIPS Open Initiative - WAVE Computing, 2024. URL: https://mips.com/products/risc-v/.
61. Contiki. Contiki-NG, 2019. URL: https://www.contiki-ng.org/.
62. Scott E. Coull, Matthew Green, and Susan Hohenberger. Access controls for oblivious and anonymous systems. In *ACM Transactions on Information and System Security - TISSEC*, volume 14, pages 1–28, 2011.
63. Ronald Cramer, Ivan Damgård, and Berry Schoenmakers. Proofs of Partial Knowledge and Simplified Design of Witness Hiding Protocols. *Advances in Cryptology - CRYPTO*, pages 174–187, 1994.
64. Quynh H. Dang. Secure Hash Standard. Technical report, 2015. doi:10.6028/NIST.FIPS.180-4.

65. Hamid Darabian, Ali Dehghantanha, Sattar Hashemi, Sajad Homayoun, and Kim-Kwang Raymond Choo. An opcode-based technique for polymorphic Internet of Things malware detection. *Concurrency and Computation: Practice and Experience*, 32, 2020.

66. Data Mining Group. PMML 4.4 - Anomaly Detection Models, 2019. URL: http://dmg.org/pmml/v4-4/AnomalyDetectionModel.html.

67. J. Deng and J. Dong. Imagenet: A large-scale hierarchical image database. *IEEE Conference on Computer Vision and Pattern Recognition - CVPR*, page 248–255, 2009.

68. Alexander W. Dent, Marc Fischlin, Mark Manulis, Martijn Stam, and Dominique Schröder. Confidential Signatures and Deterministic Signcryption. *Public Key Cryptography - PKC*, pages 462–479, 2010.

69. Ozgur Depren and Murat Topallar. An intelligent intrusion detection system (IDS) for anomaly and misuse detection in computer networks. *Expert Systems with Applications*, 29:713–722, 2005.

70. Snehal Deshmukh and S. S. Sonavane. Security protocols for Internet of Things: A survey. *International Conference On Nextgen Electronic Technologies: Silicon to Software - ICNETS2*, pages 71–74, 2017.

71. Jacob Devlin, Ming-Wei Chang, Kenton Lee, Kristina Toutanova Google, and A I Language. BERT: Pre-training of Deep Bidirectional Transformers for Language Understanding. *North American Chapter of the Association for Computational Linguistics: Human Language Technologies -NAACL-HLT*, 1:4171–4186, 2019.

72. William Diehl, Abubakr Abdulgadir, Jens-Peter Kaps, and Kris Gaj. Comparing the cost of protecting selected lightweight block ciphers against differential power analysis in low-cost FPGAs. In *International Conference on Field Programmable Technology - ICFPT*, pages 128–135, 2017.

73. Tassos Dimitriou and Ghassan O. Karame. Enabling Anonymous Authorization and Rewarding in the Smart Grid. *IEEE Transactions on Dependable and Secure Computing - TDSC*, 14:565–572, 2017.

74. Soufiene Djahel, Ronan Doolan, Gabriel-Miro Muntean, and John Murphy. A Communications-Oriented Perspective on Traffic Management Systems for Smart Cities: Challenges and Innovative Approaches. *IEEE Communications Surveys & Tutorials*, 17:125–151, 2015.

75. Ensieh Modiri Dovom, Amin Azmoodeh, Ali Dehghantanha, David Ellis Newton, Reza M. Parizi, and Hadis Karimipour. Fuzzy pattern tree for edge malware detection and categorization in IoT. *Journal of Systems Architecture*, 97:1–7, 2019.

76. Costas Efthymiou and Georgios Kalogridis. Smart Grid Privacy via Anonymization of Smart Metering Data. In *IEEE International Conference on Smart Grid Communications*, pages 238–243, 2010.

77. Taher Elgamal. A public key cryptosystem and a signature scheme based on discrete logarithms. *IEEE Transactions On Information Theory*, 31:469–472, 1985.

78. Keita Emura, Atsuko Miyaji, and Mohammad Shahriar Rahman. Dynamic attribute-based signcryption without random oracles. *International Journal of Applied Cryptography*, 2:199, 2012.

79. ETSI. European Telecommunications Standards Institute, 2021. URL: `https://www.etsi.org/`.

80. Davide Quarta Federico Maggi. When Machines Can't Talk: Security and Privacy Issues Data Protocols, 2018. URL: `https://www.blackhat.com/eu-18/briefings/schedule/index.html#when-machines-cant-talk-security-and-privacy-issues-of-machine-to-machine-data-protocols-12722`.

81. Fang Feng, Xin Liu, Binbin Yong, Rui Zhou, and Qingguo Zhou. Anomaly detection in ad-hoc networks based on deep learning model: A plug and play device. *Ad Hoc Networks*, 84:82–89, 2019.

82. FPAnalyst. Flashpoint - Attack of Things!, 2016. URL: `https://www.flashpoint-intel.com/blog/emerging-threats/attack-of-things/`.

83. Jérôme François, Humberto Abdelnur, Radu State, and Olivier Festor. Automated Behavioral Fingerprinting. In *Recent Advances in Intrusion Detection (RAID), Lecture Notes in Computer Science*, volume 5758 LNCS, pages 182–201, 2009.

84. Jason Franklin, Damon Mccoy, Parisa Tabriz, Vicentiu Neagoe, Jamie Van Randwyk, and Douglas Sicker. Passive Data Link Layer 802.11 Wireless Device Driver Fingerprinting *. In *USENIX Security Symposium - USENIX-SS*, page Article No. 12, 2006.

85. Free Software Foundation. objdump(1): info from object files - Linux man page, 2009. URL: `https://linux.die.net/man/1/objdump`.

86. Edoardo Gaetani, Leonardo Aniello, Roberto Baldoni, Federico Lombardi, Andrea Margheri, and Vladimiro Sassone. Blockchain-based database to ensure data integrity in cloud computing environments. In *CEUR Workshop*, volume 1816, pages 146–155, 2017.

87. Sanjam Garg, Craig Gentry, Shai Halevi, Amit Sahai, and Brent Waters. Attribute-based encryption for circuits from multilinear maps. *Annual Cryptology Conference - CRYPTO*, 8043 LNCS(PART 2):479–499, 2013.

88. Rosario Gennaro, Craig Gentry, and Bryan Parno. Non-interactive Verifiable Computing: Outsourcing Computation to Untrusted Workers. *Advances in Cryptology -CRYPTO*, pages 465–482, 2010.

89. Craig Gentry. Practical identity-based encryption without random oracles. *Advances in Cryptology, Lecture Notes in Computer Science - EUROCRYPT*, 4004:445–464, 2006.

90. Craig Gentry. Fully homomorphic encryption using ideal lattices. *ACM symposium on Symposium on theory of computing - STOC*, page 169, 2009.

91. Craig Gentry and Shai Halevi. Implementing Gentry's Fully-Homomorphic Encryption Scheme. *Advances in Cryptology - EUROCRYPT*, pages 129–148, 2011.

92. Google. TensorFlow. 2020. URL: `https://www.tensorflow.org/`.

93. Vipul Goyal, Omkant Pandey, Amit Sahai, and Brent Waters. Attribute-based encryption for fine-grained access control of encrypted data. *ACM conference on Computer and communications security - CCS*, page 89, 2006.

94. Matthew Green, Susan Hohenberger, and Brent Waters. Outsourcing the Decryption of ABE Ciphertexts. *USENIX conference on Security*, pages 34–34, 2011.

95. Harm Griffioen and Christian Doerr. Examining Mirai's Battle over the Internet of Things. *ACM Conference on Computer and Communications Security - CCS*, pages 743–755, 2020.

96. AHIMA Work Group. Integrity of the Healthcare Record: Best Practices for EHR Documentation (2013 update). *Journal of American Health Information Management Association - AHIMA*, 84:58–62, 2013.

97. Deepsubhra Guha Roy, Puja Das, Debashis De, and Rajkumar Buyya. QoS-aware secure transaction framework for internet of things using blockchain mechanism. *Journal of Network and Computer Applications*, 144:59–78, 2019.

98. Fuchun Guo, Yi Mu, Willy Susilo, Homer Hsing, Duncan S. Wong, and Vijay Varadharajan. Optimized identity-based encryption from bilinear pairing for lightweight devices. *IEEE Transactions on Dependable and Secure Computing*, 14:211–220, 3 2017.

99. R U I Guo, Huixian Shi, Qinglan Zhao, and Dong Zheng. Attribute-Based Signature Scheme With Multiple Authorities for Blockchain in Electronic Health Records Systems. *IEEE Access Special Section on Research Challenges and Opportunities in Security and Privacy of BlockChain Technologies*, 6:11676–11686, 2018.

100. Shaniqng Guo and Yingpei Zeng. Attribute-based signature scheme. *International Conference on Information Security and Assurance - ISA*, pages 509–511, 2008.

101. Guru99. RNN(Recurrent Neural Network) Tutorial: TensorFlow Example, 2020. URL: https://www.guru99.com/rnn-tutorial.html.

102. Stuart Haber and Benny Pinkas. Securely combining public-key cryptosystems. In *ACM conference on Computer and Communications Security - CCS*, page 215, 2001.

103. Hamed HaddadPajouh, Ali Dehghantanha, Raouf Khayami, and Kim Kwang Raymond Choo. A deep Recurrent Neural Network based approach for Internet of Things malware threat hunting. *Future Generation Computer Systems*, 85:88–96, 2018.

104. Salma A. Hamad, Dai H. Tran, Quan Z. Sheng, and Wei E. Zhang. BERTDeep-Ware: A Cross-architecture Malware Detection Solution for IoT Systems. *IEEE International Conference on Trust, Security and Privacy in Computing and Communications - TrustCom*, pages 927–934, 2021.

105. Salma Abdalla Hamad, Quan Z. Sheng, Dai Hoang Tran, Wei Emma Zhang, and Surya Nepal. A Behavioural Network Traffic Novelty Detection for the Internet of Things Infrastructures. In *International Symposium on Parallel Architectures, Algorithms and Programming - PAAP*, volume CCIS 1362, pages 174–186, 2020.

106. Salma Abdalla Hamad, Quan Z. Sheng, Wei Emma Zhang, and Surya. Nepal. Realizing an Internet of Secure Things: A Survey on Issues and Enabling Technologies. *IEEE Communications Surveys & Tutorials - COMST*, 22:1372–1391, 2020.

107. Salma Abdalla Hamad, Wei Emma Zhang, Quan Z. Sheng, and Surya Nepal. IoT device identification via network-flow based fingerprinting and learning. *IEEE International Conference on Trust, Security and Privacy in Computing and Communications - TrustCom*, pages 103–111, 2019.

108. Ayyoob Hamza, Hassan Habibi Gharakheili, and Benson. Detecting Volumetric Attacks on IoT Devices via SDN-Based Monitoring of MUD Activity. *ACM Symposium on SDN Research - SOSR*, pages 36–48, 2019.

109. Jinguang Han, Willy Susilo, Yi Mu, Jianying Zhou, and Man Ho Au. PPDCP-ABE: Privacy-Preserving Decentralized Cipher-Policy Attribute-Based Encryption. *ESORICS*, pages 73–90, 2014.

110. Mahmudul Hasan, Md. Milon Islam, Md Ishrak Islam Zarif, and M.M.A. Hashem. Attack and anomaly detection in IoT sensors in IoT sites using machine learning approaches. *IEEE Internet of Things - IoT-J*, 7:1–14, 2019.

111. José Martínez Heras and Alessandro Donati. Enhanced telemetry monitoring with novelty detection. *AI Magazine*, 35:37–46, 2014.

112. José L. Hernández-Ramos, Jorge Bernal Bernabe, M. Victoria Moreno, and Antonio F. Skarmeta. Preserving smart objects privacy through anonymous and accountable access control for a M2M-enabled internet of things. *Sensors*, 15:15611–15639, 2015.

113. Dang Hai Hoang and Ha Duong Nguyen. Detecting Anomalous Network Traffic in IoT Networks. *International Conference on Advanced Communication Technology - ICACT*, pages 1143–1152, 2019.

114. Dang Kien Hoang, Dai Tho Nguyen, and Duy Loi Vu. IoT Malware Classification Based on System Calls. *International Conference on Computing and Communication Technologies - RIVF*, pages 1–6, 2020.

115. Susan Hohenberger and Brent Waters. Online/offline attribute-based encryption. *International Workshop on Public Key Cryptography - PKC*, 8383 LNCS:293–310, 2014.

116. Xinyi Huang, Joseph K. Liu, Shaohua Tang, Yang Xiang, Kaitai Liang, Li Xu, and Jianying Zhou. Cost-effective authentic and anonymous data sharing with forward security. *IEEE Transactions on Computers*, 64(4):971–983, 2015.

117. Huggingface. Transformers — transformers 4.0.0 documentation. 2020. URL: `https://huggingface.co/transformers/`.

118. Junbeom Hur and Dong Kun Noh. Attribute-Based Access Control with Efficient Revocation in Data Outsourcing Systems. *IEEE Transactions on Parallel and Distributed Systems - CSDL*, 22:1214–1221, 2011.

119. Hidenori Ide and Takio Kurita. Improvement of learning for CNN with ReLU activation by sparse regularization. *The International Joint Conference on Neural Networks - IJCNN*, pages 2684–2691, 2017.

120. Sidra Ijaz, Munam Ali, Abid Khan, and Mansoor Ahmed. Smart Cities: A Survey on Security Concerns. *International Journal of Advanced Computer Science and Applications*, 7:1–14, 2016.

121. Markus Jakobsson and Susanne Wetzel. Secure Server-Aided Signature Generation. In Kwangjo Kim, editor, *Public Key Cryptography*, pages 383–401, Berlin, Heidelberg, 2001. Springer Berlin Heidelberg.

122. Beakcheol Jang, Myeonghwi Kim, Gaspard Harerimana, Sang-ug Kang, and Jong Wook Kim. Bi-LSTM Model to Increase Accuracy in Text Classification: Combining Word2vec CNN and Attention Mechanism. *Applied Sciences*, 10:5841, 2020.

123. Julian Jang-Jaccard and Surya Nepal. A survey of emerging threats in cybersecurity. *Journal of Computer and System Sciences*, 80:973–993, 8 2014.

124. Jueun Jeon, Jong Hyuk Park, and Young Sik Jeong. Dynamic Analysis for IoT Malware Detection with Convolution Neural Network Model. *IEEE Access*, 8:96899–96911, 2020.

125. Jeremy Norman. Kevin Ashton Invents the Term "The Internet of Things" : History of Information, 2013. URL: https://www.historyofinformation.com/detail.php?id=3411.

126. Josh Fruhlinger. Mirai botnet explained, 2018. URL: https://csoonline.com/article/3258748/the-mirai-botnet-explained-how-teen-scammers-and-cctv-cameras-almost-brought-down-the-internet.html.

127. Taeho Jung, Xiang-Yang Li, Zhiguo Wan, and Meng Wan. Privacy preserving cloud data access with multi-authorities. *IEEE INFOCOM*, pages 2625–2633, 2013.

128. Monnappa K. A. *Learning Malware Analysis: Explore the Concepts, Tools, and Techniques to Analyze and Investigate Windows Malware.* Birmingham: Packt Publishing, Limited, 2018.

129. Nesrine Kaaniche and Maryline Laurent. Attribute-based signatures for supporting anonymous certification. *European Symposium on Research in Computer - ESORICS*, 9878 LNCS:279–300, 2016.

130. Sai Praveen Kadiyala, Manaar Alam, Yash Shrivastava, Sikhar Patranabis, Muhamed Fauzi Bin Abbas, Arnab Kumar Biswas, Debdeep Mukhopadhyay, and Thambipillai Srikanthan. LAMBDA: Lightweight Assessment of Malware for emBeddeD Architectures. *ACM Transactions on Embedded Computing Systems - TECS*, 19, 2020.

131. Rosco Kalis and Adam Belloum. Validating data integrity with blockchain. *The International Conference on Cloud Computing Technology and Science - CloudCom*, pages 272–277, 2018.

132. Kaspersky Labs. Kaspersky Embedded Security Solutions Secure OS for the Internet of Things. Technical report, 2017. URL: www.securelist.com.

133. Keras. Home - Keras Documentation. 2019. URL: https://keras.io/.

134. Dalia Khader. Attribute Based Group Signature with Revocation. *IACR Cryptology ePrint Archive*, page Report 2007/241, 2007.

135. Tadayoshi Kohno and Andre Broido. Remote physical device fingerprinting. In *IEEE Transactions on Dependable and Secure Computing - TDSC*, volume 2, pages 93–108, 2005.

136. Mohit Kumar. Linux worm targeting Routers, Set-top boxes and Security Cameras with PHP-CGI Vulnerability, 2013. URL: https://thehackernews.com/2013/11/Linux-ELF-malware-php-cgi-vulnerability.html.

137. Neeraj Kumar, Rahat Iqbal, Sudip Misra, and Joel J. P.C. Rodrigues. An intelligent approach for building a secure decentralized public key infrastructure in VANET. *Journal of Computer and System Sciences*, 81:1042–1058, 2015.

138. Bouchra Lamrini, Augustin Gjini, and Simon Daudin. Anomaly Detection Using Similarity-based One-Class SVM for Network Traffic Characterization. *International Workshop on Principles of Diagnosis - DX*, pages 1–8, 2018.

139. Ruggero Lanotte, Massimo Merro, Riccardo Muradore, and Luca Vigano. A Formal Approach to Cyber-Physical Attacks. *IEEE Computer Security Foundations Symposium -CSF*, pages 436–450, 2017.

140. Eryk Lewinson. Outlier Detection with Isolation Forest - Towards Data Science, 2018. URL: https://towardsdatascience.com/outlier-detection-with-isolation-forest-3d190448d45e.

141. Allison Lewko and Brent Waters. New techniques for dual system encryption and fully secure HIBE with short ciphertexts. *Theory of Cryptography Conference - TCC*, 5978 LNCS:455–479, 2010.

142. Chun Ta Li, Tsu Yang Wu, Chin Ling Chen, Cheng Chi Lee, and Chien Ming Chen. An efficient user authentication and user anonymity scheme with provably security for IoT-based medical care system. *Sensors*, 17(7), 2017.

143. Hongwei Li, Yuanshun Dai, and Bo Yang. Identity-Based Cryptography for Cloud Security. *IACR Cryptology ePrint Archive*, 169:1–9, 2011.

144. Jie Li, Huang Lu, and Mohsen Guizani. ACPN: A novel authentication framework with conditional privacy-preservation and non-repudiation for VANETs. *IEEE Transactions on Parallel and Distributed Systems*, 26(4):938–948, 2015.

145. Jin Li, Man Ho Au, Willy Susilo, Dongqing Xie, and Kui Ren. Attribute-based signature and its applications. *ACM symposium on Information, computer and communications security - ASIACCS*, pages 60–69, 2010.

146. Jin Li, Xiaofeng Chen, Mingqiang Li, Jingwei Li, Patrick P.C. Lee, and Wenjing Lou. Secure Deduplication with Efficient and Reliable Convergent Key Management. *IEEE Transactions on Parallel and Distributed Systems*, 25:1615–1625, 2014.

147. Chao Liang, Bharanidharan Shanmugam, Sami Azam, Asif Karim, Ashraful Islam, Mazdak Zamani, Sanaz Kavianpour, and Norbik Bashah Idris. Intrusion Detection System for the Internet of Things Based on Blockchain and Multi-Agent Systems. *Electronics*, 9:1–27, 2020.

148. Kaitai Liang, Liming Fang, Willy Susilo, and Duncan S. Wong. A Ciphertext-Policy Attribute-Based Proxy Re-encryption with Chosen-Ciphertext Security. *5th International Conference on Intelligent Networking and Collaborative Systems*, pages 552–559, 2013.

149. Xiaohui Liang, Zhenfu Cao, Huang Lin, and Jun Shao. Attribute based proxy re-encryption with delegating capabilities. *International Symposium on Information, Computer, and Communications Security - ASIACCS*, page 276–286, 2009.

150. Xueping Liang, Juan Zhao, Sachin Shetty, and Danyi Li. Towards data assurance and resilience in IoT using blockchain. *IEEE Military Communications Conference - MILCOM*, pages 261–266, 2017.

151. Benoît Libert and Jean-Jacques Quisquater. Efficient Signcryption with Key Privacy from Gap Diffie-Hellman Groups. *Public Key Cryptography - PKC*, pages 187–200, 2004.

152. Wang Linlin, Lizhi Peng, Majing Su, Bo Yang, and Xiaoqing Zhou. On the Impact of Packet Inter Arrival Time for Early Stage Traffic Identification. *IEEE International Conference on Internet of Things - iThings*, pages 510–515, 2016.

153. Fei Tony Liu, Kai Ming Ting, and Zhi Hua Zhou. Isolation Forest. *IEEE International Conference on Data Mining - ICDM*, pages 413–422, 2008.

154. Fei Tony Liu, Kai Ming Ting, and Zhi Hua Zhou. Isolation-based anomaly detection. *ACM Transactions on Knowledge Discovery from Data - TKDD*, 6:1–37, 2012.

155. Hongwei Liu, Ping Zhu, Zehong Chen, Peng Zhang, and Zoe L. Jiang. Attribute-Based Encryption Scheme Supporting Decryption Outsourcing and Attribute Revocation in Cloud Storage. *IEEE International Conference on*

Computational Science and Engineering - CSE and IEEE International Conference on Embedded and Ubiquitous Computing - EUC, pages 556–561, 2017.

156. Jie Liu. Preventing leakages of business secrets from encrypt data stored in the cloud. *IEEE International Conference on High Performance Computing and Communications - HPCC- CSS- ICESS*, pages 1373–1376, 2015.

157. Yu Liu, Kai Guo, Xiangdong Huang, Zhou Zhou, and Yichi Zhang. Detecting Android Malwares with High-Efficient Hybrid Analyzing Methods. *Mobile Information Systems*, 2018.

158. Zhusong Liu, Hongyang Yan, and Zhike Li. Server-aided anonymous attribute-based authentication in cloud computing. *Future Generation Computer Systems*, 52:61–66, 2015.

159. Lucas Lundgren. Taking Over The World Through MQTT-AfterMath Black Hat USA 2017 | Briefings, 2017. URL: `https://www.blackhat.com/us-17/briefings.html#lucas-lundgren`.

160. M. Fathy M. Sabokrou, M. Khalooei and E. Adeli. Adversarially learned one-class classifier for novelty detection. *IEEE Conference on Computer Vision and Pattern Recognition - CVPR*, page 3379–3388, 2018.

161. Raj Maede Zolanvari, Jain. IoT Security: A Survey. *Recent Advances in Networking (Data Center Virtualization, SDN, Big Data, Internet of Things)*, pages 1–15, 2015.

162. Adnan Mahmood, Wei Emma Zhang, and Quan Z. Sheng. Software-Defined Heterogeneous Vehicular Networking: The Architectural Design and Open Challenges. *Future Internet*, 11:1–17, 2019.

163. Riyadh Mahmood, Nariman Mirzaei, and Sam Malek. EvoDroid: Segmented evolutionary testing of Android apps. In *ACM Symposium on the Foundations of Software Engineering - SIGSOFT*, pages 599–609, 2014.

164. Hemanta Maji, Manoj Prabhakaran, and Mike Rosulek. Attribute-Based Signatures: Achieving Attribute-Privacy and Collusion-Resistance. *IACR Cryptology ePrint Archive*, pages 1–23, 2008.

165. Hemanta K Maji, Manoj Prabhakaran, and Mike Rosulek. Attribute-Based Signatures. *Topics in Cryptology - CT-RSA*, pages 376–392, 2011.

166. Antoni Martinez-Balleste, Pablo Perez-martinez, and Agusti Solanas. The pursuit of citizens' privacy: a privacy-aware smart city is possible. *IEEE Communications Magazine*, 51:136–141, 2013.

167. José Antonio Martínez-Heras, Alessandro Donati, Marcus G.F. Kirsch, and Frederic Schmidt. New telemetry monitoring paradigm with Novelty Detection. *International Conference on Space Operations - SpaceOps*, pages 1–9, 2012.

168. Jack McBride, Budi Arief, and Julio Hernandez-Castro. Security Analysis of Contiki IoT Operating System. Technical report, 2018. URL: `http://kar.kent.ac.uk/67379/`.

169. Kerry A Mckay, Larry Bassham Meltem, Sönmez Turan, and Nicky Mouha. Report on Lightweight Cryptography. Technical report, 2017.

170. Niall McLaughlin, Adam Doupé, Gail Joon Ahn, Jesus Martinez del Rincon, BooJoong Kang, Suleiman Yerima, Paul Miller, Sakir Sezer, Yeganeh Safaei, Erik Trickel, and Ziming Zhao. Deep Android Malware Detection. In *ACM Conference on Data and Application Security and Privacy - CODASPY*, pages 301–308, 2017.

171. William Mehuron. FIPS PUB 198 The Keyed-Hash Message Authentication Code (HMAC) Category: Computer Security Cryptography. Technical report, Information Technology Laboratory National Institute of Standards and Technology Gaithersburg, MD, 2008. URL: https://nvlpubs.nist.gov/nistpubs/FIPS/NIST.FIPS.198-1.pdf.

172. Yair Meidan, Michael Bohadana, Yael Mathov, Yisroel Mirsky, Asaf Shabtai, Dominik Breitenbacher, and Yuval Elovici. N-BaIoT-Network-based detection of IoT botnet attacks using deep autoencoders. *IEEE Pervasive Computing - XPL*, 17:12–22, 2018.

173. Yair Meidan, Michael Bohadana, Asaf Shabtai, Martin Ochoa, Nils Ole Tippenhauer, Juan Davis Guarnizo, and Yuval Elovici. Detection of Unauthorized IoT Devices Using Machine Learning Techniques. *arXiv:1709.04647*, 2017.

174. Carlos Aguilar Melchor, Pierre-Louis Cayrel, and Philippe Gaborit. A New Efficient Threshold Ring Signature Scheme Based on Coding Theory. *IEEE Transaction On Information Theory*, 57:1–16, 2011.

175. Microchip. New/Popular 8-bit Microcontrollers Products - Microchip Technology Inc. URL: https://www.microchip.com/ParamChartSearch/chart.aspx?branchID=1012.

176. Markus Miettinen, Samuel Marchal, Ibbad Hafeez, N. Asokan, Ahmad-Reza Sadeghi, and Sasu Tarkoma. IoT SENTINEL: Automated Device-Type Identification for Security Enforcement in IoT. In *IEEE International Conference on Distributed Computing Systems - ICDCS*, pages 2177–2184, 2017.

177. Michael Miller. Cloud Computing : Web-Based Applications That Change the Way You Work and Collaborate Online. *Que Publishing*, pages 1–29, 2009.

178. Mahdi H. Miraz and Maaruf Ali. Blockchain enabled enhanced IoT ecosystem security. *Lecture Notes of the Institute for Computer Sciences, Social-Informatics and Telecommunications Engineering - LNICST*, 200:38–46, 2018.

179. Yisroel Mirsky, Tomer Doitshman, and Yuval Elovici. Kitsune: An Ensemble of Autoencoders for Online Network Intrusion Detection. *Network and Distributed Systems Security Symposium - NDSS*, pages 1–15, 2018.

180. Axel Moinet, Benoît Darties, and Jean-Luc Baril. Blockchain based trust & authentication for decentralized sensor networks. *IEEE Security & Privacy, Special Issue on Blockchain - SP*, pages 1–6, 2017.

181. Andrew Moore, Denis Zuev, Michael Crogan, Andrew W. Moore, and Michael L. Crogan. Discriminators for use in flow-based classification Discriminators for use in flow-based classification. Technical report, Queen Mary University of London, 2005.

182. Nour Moustafa, Benjamin Turnbull, and Kim Kwang Raymond Choo. An Ensemble Intrusion Detection Technique based on proposed Statistical Flow Features for Protecting Network Traffic of Internet of Things. *IEEE Internet of Things Journal - IoT-J*, 6:4815–4830, 2018.

183. Nour Moustafa Abdelhameed Moustafa. *Designing an online and reliable statistical anomaly detection framework for dealing with large high-speed network traffic*. PhD thesis, The University of New South Wales - UNSW, 2017.

184. Nikshepa N. and Vasudeva Pai. Survey on IoT Security Issues and Security Protocols. *International Journal of Computer Applications*, 180:975–8887, 2018.

185. Satoshi Nakamoto. Bitcoin: A Peer-to-Peer Electronic Cash System, 2008. URL: www.bitcoin.org.
186. Shivaramakrishnan Narayan, Martin Gagné, and Reihaneh Safavi-Naini. Privacy preserving EHR system using attribute-based infrastructure. In *ACM workshop on Cloud computing security workshop - CCSW*, pages 47–52, 2010.
187. Sheraz Naseer, Yasir Saleem, Shehzad Khalid, Muhammad Khawar Bashir, Jihun Han, Muhammad Munwar Iqbal, and Kijun Han. Enhanced Network Anomaly Detection Based on Deep Neural Networks. *IEEE Access*, 6:48231–48246, 2018.
188. Nataliia Neshenko, Elias Bou-Harb, Jorge Crichigno, Georges Kaddoum, and Nasir Ghani. Demystifying IoT Security: An Exhaustive Survey on IoT Vulnerabilities and a First Empirical Look on Internet-Scale IoT Exploitations. *IEEE Communications Surveys and Tutorials - COMST*, 21:2702–2733, 2019.
189. Quoc Dung Ngo, Huy Trung Nguyen, Van Hoang Le, and Doan Hieu Nguyen. A survey of IoT malware and detection methods based on static features. *ICT Express*, 6:280–286, 2020.
190. Anne H. Ngu, Mario Gutierrez, Vangelis Metsis, Surya Nepal, and Quan Z. Sheng. IoT Middleware: A Survey on Issues and Enabling Technologies. *IEEE Internet of Things Journal*, 4:1–20, 2017.
191. Kim Thuat Nguyen, Nouha Oualha, and Maryline Laurent. Securely outsourcing the ciphertext-policy attribute-based encryption. *World Wide Web*, 21:169–183, 2018.
192. Thien Duc Nguyen, Samuel Marchal, Markus Miettinen, Hossein Fereidooni, and N. Asokan. DIoT: A Federated Self-learning Anomaly Detection System for IoT. *IEEE International Conference on Distributed Computing Systems - ICDCS*, pages 1–12, 2019.
193. Thuy T.T. Nguyen and Grenville Armitage. A survey of techniques for internet traffic classification using machine learning. *IEEE Communications Surveys and Tutorials - COMST*, 10:56–76, 2008.
194. Jianbing Ni, Xiaodong Lin, Kuan Zhang, and Xuemin Shen. Privacy-Preserving Real-Time Navigation System Using Vehicular Crowdsourcing. *IEEE Vehicular Technology Conference -VTC-Fall*, pages 1–5, 9 2016.
195. Jianbing Ni, Aiqing Zhang, Xiaodong Lin, and Xuemin Sherman Shen. Security, Privacy, and Fairness in Fog-Based Vehicular Crowdsensing. *IEEE Communications Magazine*, 55:146–152, 2017.
196. Weina Niu, Rong Cao, Xiaosong Zhang, Kangyi Ding, Kaimeng Zhang, and Ting Li. OpCode-Level Function Call Graph Based Android Malware Classification Using Deep Learning. *Sensors*, 20:1–21, 2020.
197. Sven Nomm and Hayretdin Bahsi. Unsupervised Anomaly Based Botnet Detection in IoT Networks. *IEEE International Conference on Machine Learning and Applications - ICMLA*, pages 1048–1053, 2019.
198. Talal H. Noor, Quan Z. Sheng, Zakaria Maamar, and Sherali Zeadally. Managing Trust in the Cloud: State of the Art and Research Challenges. *IEEE Computer*, 49:34–45, 2016.
199. Job Noorman, Felix Freiling, Jo Van Bulck, Jan Tobias Mühlberg, Frank Piessens, Pieter Maene, Bart Preneel, Ingrid Verbauwhede, Johannes Götzfried, and Tilo Müller. Sancus 2.0: A Low-Cost Security Architecture for IoT Devices. *ACM Transactions on Privacy and Security*, 20:1–33, 2017.

200. Norton by Symantec. Zero-day vulnerability, 2019. URL: `https://us.norton.com/internetsecurity-emerging-threats-how-do-zero-day-vulnerabilities-work-30sectech.html`.

201. Rajvardhan Oak, Min Du, David Yan, Harshvardhan Takawale, and Idan Amit. Malware detection on highly imbalanced data through sequence modeling. *ACM workshop on Artificial Intelligence and Security - AISec*, pages 37–48, 2019.

202. Vanga Odelu, Ashok Kumar Das, Muhammad Khurram Khan, Kim Kwang Raymond Choo, and Minho Jo. Expressive CP-ABE scheme for mobile devices in IoT satisfying constant-size keys and ciphertexts. *IEEE Access*, 5:3273–3283, 2017.

203. Sunday Oyinlola Ogundoyin. An autonomous lightweight conditional privacy-preserving authentication scheme with provable security for vehicular ad-hoc networks. *International Journal of Computers and Applications*, pages 1–16, 2018.

204. Go Ohtake, Reihaneh Safavi-Naini, and Liang Feng Zhang. Outsourcing Scheme of ABE Encryption Secure against Malicious Adversary. In *International Conference on Information Systems Security and Privacy - ICISSP*, pages 71–82, 2017.

205. Himss Organization. 2017 HIMSS Cybersecurity survey., 2017.

206. Osman Bicer and Alptekin Kupcu. Versatile ABS: Usage Limited, Revocable, Threshold Traceable, Authority Hiding, Decentralized Attribute Based Signatures. *Cryptology ePrint Archive: Report 2019/203*, pages 1–21, 2019.

207. Poojan Oza and Vishal M. Patel. One-Class Convolutional Neural Network. *IEEE Signal Processing Letters - XPL*, 26:277–281, 2019.

208. Tapas Pandit, Sumit Kumar Pandey, and Rana Barua. Attribute-Based Signcryption : Signer Privacy, Strong Unforgeability and IND-CCA2 Security in Adaptive-Predicates Attack. *ProvSec*, 8782:274–290, 2014.

209. Sandhya Peddabachigari, Ajith Abraham, Crina Grosan, and Johnson Thomas. Modeling intrusion detection system using hybrid intelligent systems. *Journal of Network and Computer Applications*, 30:114–132, 2007.

210. Salvador Perez, Jose L. Hernandez-Ramos, Diego Pedone, Domenico Rotondi, Leonardo Straniero, and Antonio F. Skarmeta. A digital envelope approach using attribute-based encryption for secure data exchange in IoT scenarios. In *Global Internet of Things Summit - GIoTS*, pages 1–6, 2017.

211. William Peters, Ali Dehghantanha, Reza M. Parizi, and Gautam Srivastava. A Comparison of State-of-the-Art Machine Learning Models for OpCode-Based IoT Malware Detection. In *Handbook of Big Data Privacy*, pages 109–120. 2020.

212. Philippe Biondi. Scapy, 2018. URL: `https://scapy.net/`.

213. Tran Nghi Phu, Kien Hoang Dang, Dung Ngo Quoc, Nguyen Tho Dai, and Nguyen Ngoc Binh. A Novel Framework to Classify Malware in MIPS Architecture-Based IoT Devices. *Security and Communication Networks*, pages 1–13, 2019.

214. Rod Pierce. Quartiles, 2016. URL: `https://mathsisfun.com/data/quartiles.html`.

215. Marco A. F. Pimentel, David A. Clifton, and Lei Clifton. A review of novelty detection. *Signal Processing*, 99:215–249, 2014.

216. S. Pouyanfar, S. Sadiq, Y. Yan, H. Tian, and Y. Tao. A Survey on Deep Learning: Algorithms, Techniques, and Applications. *ACM Computing Surveys - CSUR*, 51:1–36, 2018.

217. Piyush Aniruddha Puranik. *Static Malware Detection using Deep Neural Networks on Static Malware Detection using Deep Neural Networks on Portable Executables Portable Executables.* PhD thesis, Nevada, Las Vegas, 2019.

218. Python Software Foundation. Python object serialization, 2020. URL: https://docs.python.org/3/library/pickle.html.

219. R. Rivest. Request for Comments: 1321 (MD5). Technical report, MIT Laboratory for computer science and RSA Data Security Inc., 1992. URL: https://www.ietf.org/rfc/rfc1321.txt.

220. Radare. Radare2. 2020. URL: https://www.radare.org/n/radare2.html.

221. Sakthi Vignesh Radhakrishnan, A. Selcuk Uluagac, and Raheem Beyah. GTID: A Technique for Physical Device and Device Type Fingerprinting. *IEEE Transactions on Dependable and Secure Computing - TDSC*, 12:519–532, 2015.

222. Radware. BrickerBot Results in PDoS (Permanent Denial of Service) Attacks, 2017. URL: https://security.radware.com/ddos-threats-attacks/brickerbot-pdos-permanent-denial-of-service/.

223. Yogachandran Rahulamathavan, Suresh Veluru, Jinguang Han, Fei Li, Muttukrishnan Rajarajan, and Rongxing Lu. User Collusion Avoidance Scheme for Privacy-Preserving Decentralized Key-Policy Attribute-Based Encryption. *IEEE Transactions on Computers*, 65:2939–2946, 2016.

224. S. Rajasegarar et al. Hyperspherical cluster based distributed anomaly detection in wireless sensor networks. *Journal of Parallel and Distributed Computing*, 74:1833– 1847, 2014.

225. Y. Sreenivasa Rao. Attribute-based online/offline signcryption scheme. *International Journal of Communication Systems*, 30:1–20, 2017.

226. Y. Sreenivasa Rao and Ratna Dutta. Efficient attribute-based signature and signcryption realizing expressive access structures. *International Journal of Information Security*, 15:81–109, 2 2016.

227. Mariana Raykova, Hang Zhao, and Steven M. Bellovin. Privacy Enhanced Access Control for Outsourced Data Sharing. *Financial Cryptography and Data Security*, pages 223–238, 2012.

228. Hex Rays. IDA Home. 2020. URL: https://www.hex-rays.com/products/idahome/.

229. Zhongru Ren, Haomin Wu, Qian Ning, Iftikhar Hussain, and Bingcai Chen. End-to-end malware detection for android IoT devices using deep learning. *Ad Hoc Networks*, 101:102098, 2020.

230. Ronald L. Rivest, Adi Shamir, and Yael Tauman. How to Leak a Secret. *Advances in Cryptology - ASIACRYPT*, pages 552–565, 2001.

231. Karen Rose, Scott Eldridge, and Lyman Chapin. The Internet of Things (IoT): An Overview | Internet Society. Technical report, 2016. URL: https://www.internetsociety.org/resources/doc/2015/iot-overview.

232. Cristina Rottondi and Giacomo Verticale. Using packet interarrival times for internet traffic classification. *IEEE Latin-American Conference on Communications, LATINCOM*, pages 1–6, 2011.

233. RSA Laboratories. What are Message Authentication Codes. URL: https://web.archive.org/web/20061020212439/,http://www.rsasecurity.com/rsalabs/node.asp?id=2177.

234. B. Rubinstein, B. Nelson, and L. Huang. Antidote: understanding and defending against poisoning of anomaly detectors. *ACM SIGCOMM conference on Internet measurement - SIGCOMM*, page 1– 14., 2009.

235. Markku-Juhani O Saarinen and Daniel Engels. A Do-It-All-Cipher for RFID: Design Requirements. *IACR Cryptology ePrint Archive*, 2012.

236. Amit Sahai and Brent Waters. Fuzzy Identity-Based Encryption. *EURO-CRYPT*, pages 457–473, 2005.

237. Mangal Sain, Young Jin Kang, and Hoon Jae Lee. Survey on security in Internet of Things: State of the art and challenges. In *International Conference on Advanced Communication Technology - ICACT*, pages 699–704, 2017.

238. Tara Salman and Deval Bhamare. Machine Learning for Anomaly Detection and Categorization in Multi-Cloud Environments. *IEEE International Conference on Cyber Security and Cloud Computing - CSCloud*, pages 97–103, 2017.

239. Mahadev Satyanarayanan. The emergence of edge computing. *Computer*, 50:30–39, 2017.

240. O. Savry, F. Pebay-Peyroula, F. Dehmas, G. Robert, and J. Reverdy. RFID Noisy Reader How to Prevent from Eavesdropping on the Communication? *Cryptographic Hardware and Embedded Systems - CHES*, pages 334–345, 2007.

241. Neetesh Saxena, Bong Jun Choi, and Rongxing Lu. Authentication and Authorization Scheme for Various User Roles and Devices in Smart Grid. *IEEE Transactions on Information Forensics and Security -XPL*, 11:907–921, 2016.

242. Scikit. Sklearn documentation, 2015. URL: https://scikit-learn.org/.

243. Scikit. Sklearn documentation, 2019. URL: https://scikit-learn.org/stable/.

244. Erwan Scornet. Random forests and kernel methods. *IEEE Transactions on Information Theory*, 62:1485–1500, 2016.

245. DI Management Services. RSA Algorithm, 2020. URL: https://www.di-mgt.com.au/rsa_alg.html.

246. Noushin Shabab. Looking for sophisticated malware in IoT devices | Securelist by Kaspersky. Technical report, 2020. URL: https://securelist.com/looking-for-sophisticated-malware-in-iot-devices/98530/.

247. Hovav Shacham and Brent Waters. Efficient Ring Signatures Without Random Oracles. In *Public Key Cryptography -PKC*, pages 166–180, 2007.

248. Trusit Shah and S. Venkatesan. Authentication of IoT Device and IoT Server Using Secure Vaults. *IEEE International Conference On Trust, Security And Privacy In Computing And Communications - TrustCom*, pages 819–824, 2018.

249. Siamak F. Shahandashti and Reihaneh Safavi-Naini. Threshold Attribute-Based Signatures and Their Application to Anonymous Credential Systems. *Progress in Cryptology - AFRICACRYPT*, pages 198–216, 2009.

250. Adi Shamir. Identity-Based Cryptosystems and Signature Schemes. *Advances in Cryptology*, pages 47–53, 1985.

251. Rohini Sharma, Ajay Guleria, and R. K. Singla. An overview of flow-based anomaly detection. *International Journal of Communication Networks and Distributed Systems - IJCNDS*, 21:220–240, 2018.

252. Shaila Sharmeen, Shamsul Huda, Jemal H. Abawajy, Walaa Nagy Ismail, and Mohammad Mehedi Hassan. Malware Threats and Detection for Industrial Mobile-IoT Networks. *IEEE Access*, 6:15941–15957, 2018. doi:10.1109/ACCESS.2018.2815660.

253. Shiqi Shen, Shruti Tople, and Prateek Saxena. AUROR: Defending Against Poisoning Attacks in Collaborative Deep Learning Systems. *Annual Conference on Computer Security Applications - ACSAC*, page 508–519, 2016.

254. Michael Sheng, Yongrui Qin, Lina Yao, and Boualem Benatallah. *Managing the Web of Things: Linking the Real World to the Web*. Morgan Kaufmann, 2017.

255. Quan Z. Sheng, Xue Li, and Sherali Zeadally. Enabling next-generation RFID applications: Solutions and challenges. *IEEE Computer*, 41, 2008.

256. Sandra Siby, Rajib Ranjan Maiti, and Nils Tippenhauer. IoTScanner: Detecting and Classifying Privacy Threats in IoT Neighborhoods. *ACM International Workshop on IoT Privacy, Trust, and Security - IoTPTS*, pages 23–30, 2017.

257. S. Sicari, A. Rizzardi, L. A. Grieco, and A. Coen-Porisini. Security, privacy and trust in Internet of things: The road ahead. *Computer Networks*, 76:146–164, 2015.

258. Simranjeet Sidhu and Bassam J. Mohd. Hardware Security in IoT Devices with Emphasis on Hardware Trojans. *Sensor and Actuator Networks*, (May 2017), 2019.

259. Federico Simmross-Wattenberg and Asensio-Pérez. Anomaly detection in network traffic based on statistical inference and {α}-stable modeling. *IEEE Transactions on Dependable and Secure Computing - TDSC*, 8:494–509, 2011.

260. Sergei P. Skorobogatov. Semi-invasive attacks-A new approach to hardware security analysis. Technical report, University of Cambridge, Computer Laboratory, 2005. URL: http://www.cl.cam.ac.uk/.

261. Tom Spring. Mirai and Hajime Locked Into IoT Botnet Battle, 2017. URL: https://threatpost.com/mirai-and-hajime-locked-into-iot-botnet-battle/125112/.

262. Statista. Global number of connected IoT devices 2015-2025 | Statista, 2020. URL: https://www.statista.com/statistics/1101442/iot-number-of-connected-devices-worldwide/.

263. Jiawei Su, Vargas Danilo Vasconcellos, Sanjiva Prasad, Sgandurra Daniele, Yaokai Feng, and Kouichi Sakurai. Lightweight Classification of IoT Malware Based on Image Recognition. *International Computer Software and Applications Conference - COMPSAC*, 2:664–669, 2018.

264. Aliya Tabassum, Aiman Erbad, and Mohsen Guizani. A survey on recent approaches in intrusion detection system in IoTs. *IEEE International Wireless Communications and Mobile Computing Conferenc - IWCMC*, pages 1190–1197, 2019.

265. Raymond K. H. Tai, Jack P. K. Ma, Yongjun Zhao, and Sherman S. M. Chow. Generic Constructions for Fully Secure Revocable Attribute-Based Encryption. In *ESORICS*, pages 532–551, 2017.

266. Hayate Takase, Ryotaro Kobayashi, Masahiko Kato, and Ren Ohmura. A prototype implementation and evaluation of the malware detection mechanism for IoT devices using the processor information. *International Journal of Information Security*, 19:71–81, 2020.

267. Tuan A. Tang, Lotfi Mhamdi, Des McLernon, Syed Ali Raza Zaidi, and Mounir Ghogho. Deep learning approach for Network Intrusion Detection in Software Defined Networking. *International Conference on Wireless Networks and Mobile Communications - WINCOM*, pages 258–263, 2016.

268. Texas Instruments. AN-453 COPS Based Automobile Instrument Cluster. URL: https://www.ti.com/lit/an/snoa744/snoa744.pdf?ts= 1710645675391.

269. Think Incredible. Brillo, Internet of Things OS - Intraway. URL: https://thinkincredible.intraway.com/blog-post/brillo-internet-of-things-os/.

270. Kai Ming Ting and Sunil Aryal. Tutorial: Which Anomaly Detector should I use? *IEEE International Conference on Data Mining - ICDM*, pages 1–143, 2018.

271. Elvis Tombini, Hervé Debar, Ludovic Mé, and Mireille Ducassé. A serial combination of anomaly and misuse IDSes applied to HTTP traffic. *Annual Computer Security Applications Conference - ACSAC*, pages 428–437, 2004.

272. Lyes Touati, Yacine Challal, and Abdelmadjid Bouabdallah. C-CP-ABE: Cooperative Ciphertext Policy Attribute-Based Encryption for the Internet of Things. *International Conference on Advanced Networking Distributed Systems and Applications*, pages 64–69, 2014.

273. Gurkan Tuna, Dimitrios G. Kogias, V. Cagri Gungor, Cengiz Gezer, Erhan Taşkın, and Erman Ayday. A survey on information security threats and solutions for Machine to Machine (M2M) communications. *Journal of Parallel and Distributed Computing*, 109:142–154, 2017.

274. Subhan Ullah, Lucio Marcenaro, Bernhard Rinner, Subhan Ullah, Lucio Marcenaro, and Bernhard Rinner. Secure Smart Cameras by Aggregate-Signcryption with Decryption Fairness for Multi-Receiver IoT Applications. *Sensors*, 19:327, 2019.

275. UNSW-Sydney. IoT Security - IoT Traffic Analysis, 2019. URL: https://iotanalytics.unsw.edu.au/.

276. Van Rijsbergen. *Information retrieval*. Butterworth-Heinemann, 2nd edition, 1979.

277. Kouliaridis Vasileios, Konstantia Barmpatsalou, Georgios Kambourakis, and Shuhong Chen. A Survey on Mobile Malware Detection Techniques. *Transactions on Information and Systems - IEICE*, E103-D(2):204–211, 2020.

278. Ashish Vaswani, Google Brain, Noam Shazeer, Niki Parmar, Jakob Uszkoreit, Llion Jones, Aidan N Gomez, Łukasz Kaiser, and Illia Polosukhin. Attention Is All You Need. *Advances in Neural Information Processing Systems - NIPS*, 30:5998–6008, 2017.

279. Ivano Verzola, Alessandro Donati, José Antonio Martínez Heras, Matthias Schubert, and Laszlo Somodi. Project sibyl: A novelty detection system for human spaceflight operations. *International Conference on Space Operations - SpaceOps*, pages 1–14, 2016.

280. Boyang Wang, Sherman S.M. Chow, Ming Li, and Hui Li. Storing Shared Data on the Cloud via Security-Mediator. *IEEE International Conference on Distributed Computing Systems*, pages 124–133, 2013.

281. Jingxuan Wang, Lucas C. K. Hui, S. M. B. Yiu, Xingmin Cui, Eric Ke Wang, and Junbin Fang. A Survey on the Cyber Attacks Against Non-linear State Estimation in Smart Grids. 1:40–56, 2016.

282. Xu Wang, Kangfeng Zheng, Xinxin Niu, Bin Wu, and Chunhua Wu. Detection of command and control in advanced persistent threat based on independent access. *IEEE International Conference on Communications - ICC*, pages 1–6, 2016.

283. WAQAS. BASHLITE malware turning millions of Linux Based IoT Devices into DDoS botnet. *Full Circle, The Independant Magazine For The Ubuntu Linux Community*, 2016. URL: https://fullcirclemagazine.org/podcasts/podcast-34/.

284. Brent Waters. Efficient Identity-Based Encryption Without Random Oracles. *International Conference on the Theory and Applications of Cryptographic Techniques – EUROCRYPT.*, 3494 LNCS:114–127, 2005.

285. Lifei Wei, Haojin Zhu, Zhenfu Cao, Xiaolei Dong, Weiwei Jia, Yunlu Chen, and Athanasios V. Vasilakos. Security and privacy for storage and computation in cloud computing. *Information Sciences*, 258:371–386, 2014.

286. Heiko Will, Kaspar Schleiser, and Jochen Schiller. A real-time kernel for wireless sensor networks employed in rescue scenarios, 2009.

287. Garsten Willems, Thorsten Holz, and Felix Freiling. Toward automated dynamic malware analysis using CWSandbox. *IEEE Security and Privacy - SP*, 5:32–39, 2007.

288. David H. Wolpert and William G. Macready. No free lunch theorems for optimization. *IEEE Transactions on Evolutionary Computation*, 1:67–82, 1997.

289. Qianhong Wu, Bo Qin, Lei Zhang, Josep Domingo-Ferrer, Oriol Farras, and Jesus A. Manjon. Contributory Broadcast Encryption with Efficient Encryption and Short Ciphertexts. *IEEE Transactions on Computers*, 65:466–479, 2016.

290. K. Q. Yan, S. C. Wang, S. S. Wang, and C. W. Liu. Hybrid Intrusion Detection System for enhancing the security of a cluster-based Wireless Sensor Network. *IEEE International Conference on Computer Science and Information Technology - ICCSIT*, pages 114–118, 2010.

291. Lei Yang, Abdulmalik Humayed, and Fengjun Li. A multi-cloud based privacy-preserving data publishing scheme for the internet of things. *The Annual Conference on Computer Security Applications - ACSAC*, pages 30–39, 2016.

292. Piyi Yang, Zhenfu Cao, and Xiaolei Dong. Fuzzy Identity Based Signature. *IACR Cryptology ePrint Archive*, pages 1–10, 2008.

293. Qiang Yang and Yang Liu. Federated Machine Learning: Concept and Applications. *ACM Transactions on Intelligent Systems and Technology - TIST*, 10:1–19, 2019.

294. Yuchen Yang, Longfei Wu, Guisheng Yin, Lijie Li, Hongbin Zhao, Guisheng Yin Lijie Li Yuchen Yang, Longfei Wu, Hongbin Zhao, Yuchen Yang, Longfei Wu, Guisheng Yin, Lijie Li, Hongbin Zhao, Guisheng Yin Lijie Li Yuchen Yang, Longfei Wu, and Hongbin Zhao. A Survey on Security and

Privacy Issues in IoT. *IEEE Internet of Things Journal*, 4:1250–1258, 2017.

295. Hongjian Yin and Leyou Zhang. Security Analysis and Improvement of An Anonymous Attribute-Based Proxy Re-encryption. *International Conference on Security, Privacy and Anonymity in Computation, Communication and Storage - SpaCCS*, pages 344–352, 2017.

296. LI Yong, Zhenyu Zeng, and Xiaofei Zhang. Outsourced decryption scheme supporting attribute revocation. *Journal of Tsinghua University (Science and Technology)*, 53:1664–1669, 2013.

297. Shucheng Yu, Cong Wang, Ren Kui, and Lou Wenjing. *Achieving secure, scalable, and fine-grained data access control in cloud computing.* 2010.

298. Zuoxia Yu, Man Ho Au, Qiuliang Xu, Rupeng Yang, and Jinguang Han. Leakage-Resilient Functional Encryption via Pair Encodings. *Australasian Conference on Information Security and Privacy*, pages 443–460, 2016.

299. Fangguo Zhang and Kwangjo Kim. ID-Based Blind Signature and Ring Signature from Pairings. *Advances in Cryptology - ASIACRYPT*, pages 533–547, 2002.

300. Jindan Zhang, Baocang Wang, Fatos Xhafa, Xu An Wang, and Cong Li. Energy-efficient secure outsourcing decryption of attribute based encryption for mobile device in cloud computation. *Journal of Ambient Intelligence and Humanized Computing*, 10:429–438, 2019.

301. Jixin Zhang, Zheng Qin, Kehuan Zhang, Hui Yin, and Jingfu Zou. Dalvik Opcode Graph Based Android Malware Variants Detection Using Global Topology Features. *IEEE Access*, 6:51964–51974, 2018.

302. Kai Zhang, Wangmeng Zuo, Yunjin Chen, Deyu Meng, and Lei Zhang. Beyond a Gaussian denoiser: Residual learning of deep CNN for image denoising. *IEEE Transactions on Image Processing*, 26:3142–3155, 2017.

303. Wei Emma Zhang, Quan Z. Sheng, Adnan Mahmood, Dai Hoang Tran, Munazza Zaib, Salma Abdalla Hamad, Abdulwahab Aljubairy, Ahoud Abdulrahmn, F. Alhazmi, Subhash Sagar, and Congbo Ma. The 10 Research Topics in the Internet of Things. *IEEE International Conference on Collaboration and Internet Computing - CIC*, pages 34–43, 2020.

304. Yinghui Zhang, Xiaofeng Chen, Jin Li, Duncan S. Wong, and Hui Li. Anonymous attribute-based encryption supporting efficient decryption test. *ACM SIGSAC symposium on Information, computer and communications security - ASIA CCS*, pages 511–516, 2013.

305. Yinghui Zhang, Jin Li, Xiaofeng Chen, and Hui Li. Anonymous attribute-based proxy re-encryption for access control in cloud computing. *Networks*, 9:2397–2411, 2016.

306. Zhen-Jiang Zhang, Chin-Feng Lai, and Han-Chieh Chao. A green data transmission mechanism for wireless multimedia sensor networks using information fusion. *IEEE Wireless Communications*, 21:14–19, 2014.

307. Zhiyong Zhang, Cheng Li, Brij B. Gupta, and Danmei Niu. Efficient Compressed Ciphertext Length Scheme Using Multi-Authority CP-ABE for Hierarchical Attributes. *IEEE Access*, 6:38273–38284, 2018.

308. Zijun Zhang. Improved Adam Optimizer for Deep Neural Networks. *IEEE/ACM International Symposium on Quality of Service - IWQoS*, pages 1–2, 2018.

309. Yuliang Zheng. Digital signcryption or how to achieve cost(signature & encryption) { cost(signature) + cost(encryption)}. *Advances in Cryptology - CRYPTO*, pages 165–179, 1997.

310. Jun Zhou, Zhenfu Cao, Xiaolei Dong, and Athanasios V. Vasilakos. Security and Privacy for Cloud-Based IoT: Challenges. *IEEE Communications Magazine*, 55:26–33, 2017.

311. Jun Zhou, Xiaodong Lin, Xiaolei Dong, and Zhenfu Cao. PSMPA: Patient self-controllable and multi-level privacy-preserving cooperative authentication in distributed m-healthcare cloud computing system. *IEEE Transactions on Parallel and Distributed Systems*, 26:1693–1703, 2015.

312. Zhibin Zhou and Dijiang Huang. Efficient and secure data storage operations for mobile cloud computing. *International Conference on Network and Service Management - CNSM*, pages 37–45, 2012.

313. Jan Henrik Ziegeldorf, Oscar Garcia Morchon, and Klaus Wehrle. Privacy in the Internet of Things: Threats and Challanges. *Security and Communication Networks*, 7:2728–2742, 2014.

Index

131

Printed in the United States
by Baker & Taylor Publisher Services